T0215337

Explorations in Place Attachment

The book explores the unique contribution that geographers make to the concept of place attachment, and related ideas of place identity and sense of place. It presents six types of places to which people become attached and provides a global range of empirical case studies to illustrate the theoretical foundations. The book reveals that the types of places to which people bond are not discrete. Rather, a holistic approach, one that seeks to understand the interactive and reinforcing qualities between people and places, is most effective in advancing our understanding of place attachment.

Jeffrey S. Smith is an Associate Professor of Geography at Kansas State University, USA.

Routledge Research in Culture, Space and Identity
Series editor:
Dr. Jon Anderson
School of Planning and Geography, Cardiff University, UK

The *Routledge Research in Culture, Space and Identity Series* offers a forum for original and innovative research within cultural geography and connected fields. Titles within the series are empirically and theoretically informed and explore a range of dynamic and captivating topics. This series provides a forum for cutting edge research and new theoretical perspectives that reflect the wealth of research currently being undertaken. This series is aimed at upper-level undergraduates, research students and academics, appealing to geographers as well as the broader social sciences, arts and humanities.

For a full list of titles in this series, please visit www.routledge.com/Routledge-Research-in-Culture-Space-and-Identity/book-series/CSI

Memory, Place and Identity
Commemoration and Remembrance of War and Conflict
Edited by Danielle Drozdzewski, Sarah De Nardi and Emma Waterton

Surfing Spaces
Jon Anderson

Violence in Place, Cultural and Environmental Wounding
Amanda Kearney

Arts in Place
The Arts, the Urban and Social Practice
Cara Courage

Explorations in Place Attachment
Edited by Jeffrey S. Smith

Explorations in Place Attachment

Edited by
Jeffrey S. Smith

LONDON AND NEW YORK

First published 2018
by Routledge

2 Park Square, Milton Park, Abingdon, Oxfordshire OX14 4RN
52 Vanderbilt Avenue, New York, NY 10017

Routledge is an imprint of the Taylor & Francis Group, an informa business

First issued in paperback 2019

Copyright © 2018 selection and editorial matter, Jeffrey S. Smith; individual chapters, the contributors

The right of Jeffrey S. Smith to be identified as the author of the editorial material, and of the authors for their individual chapters, has been asserted in accordance with sections 77 and 78 of the Copyright, Designs and Patents Act 1988.

All rights reserved. No part of this book may be reprinted or reproduced or utilised in any form or by any electronic, mechanical, or other means, now known or hereafter invented, including photocopying and recording, or in any information storage or retrieval system, without permission in writing from the publishers.

Notice:
Product or corporate names may be trademarks or registered trademarks, and are used only for identification and explanation without intent to infringe.

British Library Cataloguing-in-Publication Data
A catalogue record for this book is available from the British Library

Library of Congress Cataloging-in-Publication Data
A catalog record for this book has been requested

ISBN: 978-1-138-72974-2 (hbk)
ISBN: 978-0-367-88712-4 (pbk)

Typeset in Times New Roman
by Apex CoVantage, LLC

To my mentors who inspired me and molded my thinking (Daniel Arreola, Richard Nostrand, Stuart Givens, Charles Collins) and to my wonderful, loving wife, Kim

Contents

Figures

Tables

Contributors

Paul C. Adams is Professor in the Department of Geography and the Environment at the University of Texas, Austin. He holds a Ph.D. from the University of Wisconsin, Madison. His research interests include geography of media and communication, representations of places, landscapes and environments, critical geopolitics, and agency and identity.

Engrid Barnett earned her Ph.D. in cultural geography from the University of Nevada, Reno. Her research strives to provide a more holistic interpretation of the history of the 1960s within the context of cultural geography. She teaches geography, humanities, and French courses at the University of Nevada, Reno.

Geoffrey L. Buckley is Professor of Geography at Ohio University. He holds a Master's degree from the University of Oregon and a Ph.D. from the University of Maryland. His research interests include historical geography, public lands, urban sustainability, environmental justice, and the evolution of mining landscapes. He is co-editor of *North American Odyssey: Historical Geographies for the Twenty-first Century* (2014).

Hélène B. Ducros earned her J.D. and Ph.D. in geography from the University of North Carolina at Chapel Hill. She is interested in heritage landscapes and the role of memory, identity, culture, and place in local development.

Douglas A. Hurt is Assistant Teaching Professor and Director of Undergraduate Studies in the Department of Geography at the University of Missouri. His research interests include geographic patterns resulting from attachment to place, public memory, and heritage tourism, as well as sport and regional identity.

Christine K. Johnson is affiliated with the University of Nevada, Reno, teaching in the departments of anthropology and geography. She holds a Ph.D. in cultural geography and a Master's in anthropology from the University of Nevada. Her primary research interest is place image, with additional interests in Pacific history and indigenous cultures.

Kimberly E. Klockow is a UCAR Research Scientist I and Policy Advisor at the NOAA Office of Weather and Air Quality in Norman, Oklahoma. She earned her Ph.D. in geography from the University of Oklahoma. Her research interests center on the intersections of weather and climate risk with risk perception, place attachment, spatial cognition, and cartographic representations of risk.

Tyra A. Olstad, an alumna of Dartmouth College, the University of Wyoming, and Kansas State University, is currently an Assistant Professor of Geography and Environmental Sustainability at SUNY Oneonta. Her research draws on her experience as a park ranger, physical science technician, and summit steward at numerous Parks and Forests throughout the United States, where she explores the intersections of landscape perception, place attachment, land management, and wilderness.

Randy A. Peppler is the director of a meteorological institute and a lecturer in geography and environmental sustainability at the University of Oklahoma. He is interested in placed-based knowledge formation and risk perception.

Chris W. Post is an Associate Professor of Geography at Kent State University at Stark. His research focuses on the heritage of place, particularly as it becomes manifest on the cultural landscape.

Rex "RJ" Rowley is an Associate Professor of Geography at Illinois State University. He has research interests in sense of place, urban geography, and geographic information science. He makes regular trips back to Kesennuma, Japan, with his students as part of a study abroad field class.

Steven M. Schnell is a Professor in the Department of Geography at Kutztown University in Kutztown, Pennsylvania. His research focuses on the myriad ways that people create attachments to place. He is currently editor of the *Journal of Cultural Geography*.

Jeffrey S. Smith is an Associate Professor of Geography at Kansas State University. His research focuses on cultural change, landscape analysis, and place attachment.

Michael Strong is an Assistant Professor at Glendale Community College in Phoenix, Arizona. His research interests address the impact that rapidly changing environments have on place attachment. He has worked on projects exploring place attachment at several sites in Mozambique.

Yolonda Youngs is an Assistant Professor at Idaho State University. She has a Ph.D. in geography from Arizona State University. Her research specialties are environmental historical geography, cultural landscapes, visual media, tourism and outdoor recreation, and national parks and protected areas.

Introduction

Putting place back in place attachment research

Jeffrey S. Smith

Humans are curious creatures; sometimes the emotional ties they make to a place defy logic. What compels people to rebuild their home in the same location that was destroyed by a devastating natural disaster? Why do people remain rooted in small, rural towns with dwindling populations and limited employment opportunities? Why do people return to the same vacation spot annually when there are boundless opportunities elsewhere? What compels people to become so emotionally connected to a place? How can the geographic perspective enhance our understanding of why people develop such intense feelings toward a place? These are the type of inquiries this book seeks to answer. It provides an innovative examination of the types of places to which people become emotionally attached.

In recent years decision makers have begun seeking ways to improve overall quality of life by building upon people's connection with place. Government officials are searching for ways to develop more resident-friendly communities, programs that might help youth of all religious and cultural backgrounds as well as the elderly feel more connected to the place they live. Administrators at assisted care facilities are looking for methods to foster a greater sense of community among retirement-aged residents. International aid organizations are experimenting with various ways to help displaced persons feel more at home in refugee camps. For most people life takes on more meaning and individuals feel more content when they develop connections to a place; *place matters*.

There is a long tradition among North American geographers who have sought a richer understanding of the character and qualities of place(s). Over 50 years ago, at a time when the discipline was searching for its identity, William Pattison (1964) identified that understanding the nature of places with all of their unique qualities, as well as how human interact with those places, bind geographers together. In his 1985 Presidential Address to the convocation of the *Association of American Geographers*, Peirce Lewis (1985) encouraged geographers to continue seeking a deeper understanding of the character of places and share that knowledge with others through thick description. A lot has changed within geography since Lewis's address. Today geography is in an age where a myopic obsession with geospatial techniques (e.g., remote sensing, GIS, geovisualization) drives the discipline. Yet, one of the things that continues to hold geography and geographers together is a desire to learn about places and interpret how people interact with those places (Hansen 2008; Murphy 2014).

Place is defined as "a meaningful location, spaces that people are attached to in one way or another" (Cresswell 2004, 7). At its root, place is a location where human activity unfolds (Wright 1947). Place is both physical and social and requires human action to define its significance (Casey 2001; Young 2001). Abstract space becomes meaningful as we modify our surrounding environment in a manner that resonates with our soul, our inner being. Geographer John Agnew's (1987) seminal work identifies three fundamental aspects of meaningful locations. *Location* refers to the fact that every place is located somewhere either absolutely (e.g., longitude and latitude) or relatively (in relation to other places). Second, every place has a unique set of characteristics or its setting sets it apart from other places. In other words, every place has a *locale*. Third, and most important to this book, is that every place is imbued with a *sense of place*, that is, how we interact with places and the emotional connections we develop with a place.

The need to belong to other people as well as places is a universal constant and is central to the human experience. As French philosopher, Simone Weil said, "[T]o be rooted in a place is the most important and least recognized need of the human soul" (Weil 1952, 43). Bonding with place grounds us by connecting us to the past, situating us in a larger social or physical environment, and helps shape our future interactions. "To be human is to live in a world that is filled with significant places: to be human is to have and know your place" (Relph 1976, 1). I assert that the focus on understanding people's connection to place falls under the umbrella of *Place Attachment*.

Cognate perspectives

Place attachment is defined as the emotional bond that develops between a person and a place. Because the concept is so complex involving psychological and sociological as well as geographic aspects, the term is often misunderstood and imprecisely defined. To make matters worse, the methods of studying place attachment vary drastically across disciplines. Some scholars approach their research from a phenomenological perspective striving to interpret how people interact with a place and the meaning it holds (Seamon and Sowers 2008; Seamon 2014). By comparison others take a more positivist, quantitative approach seeking to measure degrees of emotional connection with little, if any, attention given to the qualities of place. General consensus among scholars determines that place attachment commonly "involves the elaborate interplay of emotion, cognition, and behavior in reference to place" (Ponzetti 2003, 1).

Scholars grounded in a spectrum of disciplinary backgrounds including anthropology, art history, cultural studies, community planning, family studies, geography, gerontology, philosophy, psychology, sociology, and tourism studies provide diverse frameworks for understanding place attachment (Low and Altman 1992; Ponzetti 2003). Over the past 40 years, scholarship on place attachment has proliferated with no less than 400 publications appearing in more than 120 different journals (Lewicka 2011). Across these various disciplines a multitude of terms and definitions are used to refer to the emotional connection people develop for a place including: *emotional ties to place* (Chamlee-Wright and Storr 2009; Lewicka 2005; Mattila 2001; van der Graaf 2008), *homeland* (Nostrand 1992; Nostrand and Estaville 2001; Smith and White 2004), *insideness* (Ponzetti 2003; Relph 1976; Rowles 1990; Seamon 2008), *place dependence* (Hernández *et al.* 2014; Jorgensen and Stedman 2006; Pretty *et al.* 2003; Raymond *et al.* 2010), *place identity* (Blake 1999; Chow and Healy 2008; Hernández *et al.* 2007; Marsh 1987; Proshansky *et al.* 1983), *rootedness* (McAndrew 1998; Tuan 1980), *sense of community* (Nasar and Julian 1995; Pretty *et al.* 2003; Perkins and Long 2002), *sense of place* (de Wit 2013; Hays 1998; Jackson 1994; Lewis 1979; Post 2008), and *topophilia/love of place* (Francaviglia 2003; Tuan 1974). "The diversity of definitions . . . reflect the different emphasis that various fields put on specific components of place and place attachment" (Manzo and Devine-Wright 2014, 2). For example, the term *place identity* is used in different ways depending on one's disciplinary background (Seamon 2014). Psychologists use place identity to refer to how people see themselves in a particular place (e.g., Mihaylov and Perkins 2014) whereas geographers use the same term to refer to the character and qualities of a place (e.g., Schnell 2003; Shortridge 1989; Wishart 2013). Because terminology varies, confusion, miscommunication, debate, and sometimes competition has prevailed with little cross-pollination as few scholars engage in discourse that transcends disciplinary boundaries (Casey 2001; Entrikin 2001).

This book seeks to break down disciplinary boundaries and contribute to the literature in three significant ways. First, it enhances our understanding of the *place* component within place attachment research by providing a typology that showcases six types of places to which people become attached. Opportunities abound for scholars to examine these six types of places from a variety of disciplinary perspectives. Second, among the numerous reviews of place attachment research (e.g., Lewicka 2011), the work by geographers has been largely ignored. This volume helps rectify that oversight by drawing attention to the extensive body of work by past and present geographers. Because place matters, the geographic perspective provides valuable insight that should not be overlooked. Third, it is intended that the book will spur further discussion and research on place attachment both within geography and across cognate fields. The spatially grounded questions asked by geographers should provide a fresh and innovative perspective for a variety of scholars seeking to better understand the place attachment phenomenon. Building upon the geographic perspective, it is easy to envision a wide variety of real-world applications for place attachment research including adaptations to global climate change, environmental migration, risk perception and natural hazards response, land use conflicts, resource management, urban renewal, community design, and social housing.

Five decades of research on place attachment

Over the course of five decades, research on place attachment has grown by leaps and bounds. The first recognized work to focus on place attachment was by psychologist Marc Fried (1963) who looked at the impact of urban renewal on Boston's

West End. He found that displaced residents were so attached to their neighborhood that when they were forced to move, the loss they felt was akin to losing a loved one. In the 1970s humanistic geographers picked up the banner and led the charge to better understand people's emotional connection to place. Over the course of four years, Yi-Fu Tuan published three seminal works that made significant advances in the field. Tuan's breakthrough book, *Topophilia* (love of place), focuses on the values, attitudes, and perceptions that people have for places (Tuan 1974). In 1976 his book chapter titled *Geopiety* provides an innovative exploration of the intersection between a person's devotion to place and their native land or country (Tuan 1976). Then, in 1977 Tuan solidified his concepts in the book *Space and Place* where he writes that place is "an archive of fond memories and splendid achievements that inspire the present; place is permanent and hence reassuring to man," and "the more ties there are to a place, the stronger is the emotional bond" (Tuan 1977, 154, 158). In 1976, one of Tuan's contemporaries, Edward Relph, published *Place and Placelessness*. From a phenomenological approach, Relph's work seeks to understand the continuum of feelings and connections (from insideness to outsideness) that people have for places (Relph 1976). The book still informs many disciplines including geography, architecture, and landscape design because it expands our vocabulary and concepts regarding people's emotional connection to place (Seamon and Sowers 2008). Together, Tuan and Relph broke new ground and significantly advanced our understanding of people's emotional relationship with place(s). Because of their work, the discipline of geography had arguably become the home of place attachment research.

In the 1980s cultural geographers' perspective on the study of place took a new turn. Critics asserted that research on place had become focused on fixed and static locations, overlooking the dynamic and evolving qualities of places. Furthermore, the main thrust of research was too concerned with how individuals interact with a place and neglected how groups of people define a place. Social theorists including John Agnew, Michael Dear, Mona Domosh, Doreen Massey, Don Mitchell, Robert Sack, Edward Soja, and Nigel Thrift helped shape a new direction on place research, one that advanced an understanding of how social and economic forces shape a place and in turn how that place shapes people's lives (Sack 1988; Young 2001).

Filling the void left by geographers, environmental psychologists took over the main thrust of research on place attachment. The 1980s were largely a definitional era where various concepts were presented, discussed, and debated. Some of the key research was advanced by Irwin Altman (1975) whose work examined how individuals and groups carve out territories for themselves. Numerous publications by Harrold Proshansky and his colleagues (1978, 1983) focused on how people self-identify in a particular place. Shumaker and Taylor (1983) is one of numerous works that seek to identify multidimensional explanations for peoples' attachment to place (i.e., demographic, economic, race/ethnicity).

In more recent decades environmental psychologists have continued to spearhead research on place attachment. In 1992 Altman and Low edited a book by which all subsequent publications on place attachment are compared. Eighteen years later Scannell and Gifford (2010) advanced the field again by providing accessible definitions and creating the first model for place attachment research (see discussion as follows). In 2011, Maria Lewicka conducted the most complete review summarizing the entire field of place attachment over the last 40 years. Lynne Manzo and Patrick Devine-Wright (both psychologists by training) published the most recent major work on place attachment (2014). Their edited book contains 15 chapters that are divided into three sections (theory, methods, and applications). Despite being advertised as offering the most current understanding about place attachment, the book provides a skewed perspective on the vast body of research because most of the chapters reflect a preoccupation with predictors of attachment that seek to quantify the strength of bonds people have for a place. Only one chapter is written by a geographer. Moreover, David Seamon's chapter is the only one that focuses on advancing our understanding of place attachment from a humanistic, phenomenological perspective. Despite their pivotal role in the 1970s, in more recent decades the work of cultural-historical geographers has been largely overlooked. The overall focus on place attachment research has shifted from a concern with understanding the qualities of place to measuring the emotional intensity people have for places (Hernández *et al*. 2014; Lewicka 2011; Williams 2014). It is clear that the field has indeed expanded beyond the emergent stage to an application stage (Scannell and Gifford 2014), but disciplinary lines are still rarely crossed.

Person, place, and process

In 2010 Leila Scannell and Robert Gifford published "Defining Place Attachment: A Tripartite Organizational Framework." The article is one of the most revolutionary works in the field because it synthesizes the vast place attachment research into a manageable model comprised of three key aspects (Person, Place, and Process). The *Person* component has mainly been the domain of social psychologists, sociologists, and anthropologists. It focuses on *who* is attached to a place. Research started out by looking at the different characteristics of individuals, but more recently began examining how groups of people bond with a particular place. One of the key underlying concepts is *sense of community* – groups of people develop an attachment to place through shared symbolic meaning among members of the group.

The *Place* component has largely been (and remains) the domain of geographers. This component focuses on to *what* people are attached and varies across scale, from the microscale (e.g., room in a house) to macro-scale (e.g., city or country). Places can be both social as well as physical in nature. Psychologists have equated social places with degrees of bonding and physical places with rootedness (Scannell and Gifford 2010). By comparison, geographers commonly look at what makes each place unique. For example, Los Angeles with its bright lights, ethnic diversity, and Hollywood culture, as well as its amenable Mediterranean climate is a good example of both a social place and a physical place to which people have strong feelings of connection. Cultural-historical geographers typically focus on the socially constructed spaces and seek to understand how they give meaning to the lives of people who dwell there. Psychologists have become increasingly interested in the Place component, but their research tends to focus on identifying what belongs in a place as well as what elements do not fit in a place (e.g., Devine-Wright 2009)

The *Process* component has been the overwhelming preoccupation of environmental psychologists, especially in Europe. The focus is on how a place becomes meaningful and what impact a place has on individuals; how does the mind react to being in a place? Commonly employing either single or multidimensional psychometric questionnaires, their main goal is to measure degrees of attached feelings and find correlations or predictive variables. Length of time in residence is the most common variable. Scannell and Gifford (2010) identify three subparts to the process component (*affect* – emotional impact, *cognition* – memories and thoughts that people have, and *behavior* – how people react physically to a place). Some other emerging concepts being explored include *proximity maintaining* (individuals want to be close to a place they love), *homesickness* (feelings of detachment), and *reconstruction of places* (bringing elements of a place people love to their current location). The newest line of research pursued by environmental psychologists examines the *expressive* (or symbolic) meaning of places to which people are attached.

According to Maria Lewicka (2011), the *person* and *process* components have received the most scrutiny among scholars. By comparison, despite being the most important dimension of place attachment (Scannell and Gifford 2010), *place* has received the least amount of attention (Lewicka 2011). As Bernardo Hernández and his colleagues write, research on place attachment "should progress from analyzing what and how much, to analyzing other questions such as how, where, when, and why." (Hernández *et al.* 2014, 134). The emphasis placed on understanding differences among individuals has probably inhibited the development of place attachment theory (Lewicka 2011).

Typology of places

I assert that the work by geographers (past and present) has much to offer place attachment research, and geographers are well equipped to shed more light on the place component because they have the inherent perspective and tools needed to look at the qualities of places. From an interrogation of the literature (especially geography) as it intersects with the concepts of place and place attachment and drawing upon concepts found in other disciplines, I have identified six types of places to which people become attached. These six places comprise a typology of places where specific human activity lends itself for the emotional bonding with that place. To date, I am unaware of any such typology. These six types of places are not exhaustive nor are they mutually exclusive. Although each chapter in this book is directly linked to one of the six types, it will become evident (especially in the *epilogue*) that they also indirectly support and reinforce other categories of places. This illustrates that place is a complex idea and there is considerable overlap in the types of places to which people become attached. The remainder of this introductory chapter explains the six types of places, identifies seminal works previously published, and highlights how each of the chapters in this book add clarity, through empirical examples, to the typology. The critical *epilogue* identifies common themes that transcend chapters and reinforce geography's contribution to place attachment research.

Secure places

Of the six types of places, this is the most intuitive. Secure places are locations where people attach deep emotional meaning because they feel safe and secure in that setting; they are womb-like places. *Home* is the most common example of a secure place. With notable exceptions, most people feel the strongest attachment to home because they tend to feel at ease and protected from risk and danger.

From scientific research to product advertising, numerous examples illustrate the importance that "home" plays in people's emotional psyche. Clare Marcus's (1995) exploration of the meaning of home is one such example. Another is Starbuck's 2016 advertising campaign that encourages people to drink Starbuck's coffee because it gives you a sense of home no matter where you travel. Likewise, among the long-time Hispano residents of the upper Rio Grande region, the term *La Querencia* is used to effectively capture the essence of secure places. *Querencia* comes from the Castilian (Old) Spanish word *querer* (to love or want) and it refers to the place of your heart's desire, the central place that anchors you, the place where you feel completely at home, where you belong (Fauntleroy 1997).

Secure places are found at different scales – individual scale, small group scale, and community scale. At the macro-scale the *homeland* concept is particularly relevant. As Nostrand and Estaville (1993) articulate, one of the key ingredients needed for the development of a homeland is a deep emotional connection to place, a place you are willing to defend with your life. If we consider secure places from a completely different perspective, a prison might be considered a secure place because it offers a stable and predictable environment for inmates.

Part I of this book explores secure places from two distinct perspectives. In Chapter 1, Michael Strong explains how residents' shared memories of home and strong connections to place eased their transition to a new settlement in Mozambique. Chapter 2 by Randy Peppler, Kim Klockow, and Richard Smith looks at perceptions of tornado risk and how people's false sense of security at home influences their outlook toward risk management.

Socializing places

From a post-structuralist perspective, Doreen Massey (2005) argues that places are dynamic, individualistic, and relational. People's identity and emotional connections are shaped by the interpersonal relations that unfold in a particular place and time. Socializing places are locations with a strong sense of community or places where people feel welcome in a social environment. The work by Ray Oldenburg

(1989) on *Third Places* is an excellent example of socializing places. Oldenburg's research tells us that our primary place is our home and people's secondary place is found at work; the place where we traditionally spend most of our time during the week. Third places are where we socialize (e.g., cafes, churches, bars, beauty salons, night-clubs). One of the most effective ways in which people develop a strong attachment to place is through the interaction they have with others in that place (Barcus and Brunn 2009; Milligan 1998).

In communities throughout Spain and Latin America the term *resolana* nicely captures the essence of a socializing place. *Resolana* refers to a spot where members of the community gather to while away the day discussing current events (Romero 2001). A popular place to gather is on the central plaza sitting on benches under shady vegetation within sight of the kiosk (bandstand). The central plaza is an inviting example of a socializing place that lies at the heart of the community, a place where people meet to affirm their membership within the community (Smith 2004).

Part II of this book begins with Hélène Ducros's exploration of rural French villages and how the built environment in cultural heritage sites is shaped to foster feelings of place attachment among locals and visitors. In Chapter 4, Jeffrey Smith examines how residents of an informal neighborhood (squatter settlement) have dispelled the area's crime-infested reputation and resisted the government's efforts at redevelopment by demonstrating that the neighborhood has a strong sense of community.

Transformative places

Transformative places are where significant events took place in a person's lifetime or places of personal growth and achievement. A different way to think about transformative places is that they are key locations that help us tell the story of our life. Within environmental psychology this line of research falls within the purview of environmental autobiography where the goal is to understand the importance of places over the course of a person's lifetime (Rivlin 1982).

In numerous small towns across the country, former residents (especially retirees) are returning to their hometown because it was an important, transformative place in their life (Howell 1983; Rowles 1990; Rubenstein and Parmelee 1992). This is but one example of a larger body of literature that seeks to explain people's desire to physically connect with places rooted in the past or grounded in one's memories (Donohoe 2014; Hoelscher and Alderman 2004; Lowenthal 2015). Not only is home (and one's hometown) an example of a transformative place, but so are venues for athletic or artistic events. To many young males a football field or a baseball diamond was the site where they experienced personal achievement leading to a transformation in their life. Bruce Springsteen's 1984 song "Glory Days" captures the essence of transformative places. Other venues could be the site of a dance recital or stage performance. If we extend the idea of transformative places, a battlefield in war could be another example. Many veterans of World War II are returning to Normandy Beach because it was a transformative place in their life. Jamie Gillen (2014) writes about the growth of tourism in Vietnam as a part of the American/Vietnam War. Gettysburg, Pennsylvania might be another transformative place, not because veterans are returning there, but because the place was instrumental in the transformation of American society during the Civil War.

Part III of this book begins with Chris Post's look at the memorialized landscapes of three tragic events and shows how visitor empathy and corporeal interaction leads to feelings of place attachment. Chapter 6 features Steven Schnell's truly innovative look at how a graphic novel depicts place, place attachment, and place alienation among immigrants.

Restorative places

As early as the 1980s the Japanese government began encouraging its citizens to take strolls in the woods to promote better health, a practice known as *shinrin-yoku* (forest bathing) (Sifferlin 2016). Not only is spending time in nature associated with certain health benefits (e.g., lower blood pressure and reduce the risks of some diseases), but researchers have found that hiking in a forested area induces positive physiological reactions (Meade and Emch 2000). Restorative places are locations that kindle a healthy spirit or quiet the mind. Psychologists have found that humans become emotionally attached to wilderness areas including seaside and mountainous locations because we experience serene feelings there (Williams 2014).

This category of place originates from the work of William Wyckoff and Lary Dilsaver (1995), which focused on the restorative qualities of the Mountainous West. Mountains are an example of restorative places as an outgrowth of the nineteenth-century romanticism which envisioned many western mountain settings as healthy and idyllic retreats from the increasing urbanized and fast-paced world beyond (Wyckoff and Dilsaver 1995). Much of Kevin Blake's work on mountain symbolism fits nicely within this category of places to which we become attached (Blake 1999, 2008, 2010). Beyond mountains, to what other natural areas do we feel emotionally connected? Do people become attached to houses of worship for the same restorative qualities? Are national parks (e.g., Grand Canyon N.P. and Yosemite N.P.) so overrun with visitors that they have lost their emotional appeal and restorative qualities?

Part IV of this book features two chapters that speak to the attached feelings people develop for wilderness areas. Chapter 7 by Yolonda Youngs focuses on the perspective of outdoor tour guides in Grand Teton National Park, while in Chapter 8 Tyra Olstad explores how rangers (with the help of outdoor enthusiasts) in the Adirondack Mountains of upstate New York manage the state park to preserve it as a "wilderness."

Validating places

There are places where human activity is infused with shared meaning. As discussed above, some of those places center around a social function. In other cases, there are places that serve as memorials to some past event, tragic or otherwise (e.g., Tiananmen Square, New York City's World Trade Center; Alfred P. Murrah building in Oklahoma City) (see also: Foote 1997). Validating places are locations where personal and group identity is reinforced. By participating and investing ourselves in a place, participants come to feel a part of that place and associate their identity with that place (Seamon 2014). Some people find the setting of a class reunion to be a validating place. Another is Mecca, Saudi Arabia where the religious beliefs of over two million Muslim people are validated.

Countless sporting venues worldwide also serve as the locus of validation. After the home team has won, people feel like their lives have meaning and what they value has meaning; an aspect of their lives, even if only temporarily, has been affirmed. It comes as no surprise that people then project those feelings by displaying their team's logo on their personal property; the logo becomes the outward expression of the strong feelings of attachment that people develop for a place that legitimizes some aspect of their life. Another aspect of validating places is generational ties to the same place. In Peru, descendants of the Inca display the skulls of their ancestors in a prominent place in the house (e.g., fireplace mantel) as a way to demonstrate their property rights. At the same time, knowing that generations of one's ancestors lived and breathed in the same place validates one's life and contributes to a strong attachment to that place.

Part V of this book begins with Douglas Hurt's examination of how the progression of three baseball stadiums in St. Louis, Missouri reflects changes in both the urban morphology and social psyche. He also assesses the role those stadia play in fostering (or hindering) feelings of place attachment among city residents. In Chapter 10 Engrid Barnett explores the quasi-ghost town of Virginia City, Nevada to illustrate how the avant-garde, inspired by the Wild West myth, have carved out a home and developed a unique place attachment. Then, Christine Johnson (Chapter 11) follows in the wake of the *HMS Bounty* to showcase how a paradisical island in the Pacific (Pitcairn) has suffered a terrible reputation, yet its residents maintain a strong attachment to place.

Vanishing places

One of the most exciting areas within place attachment research looks at the dynamic qualities (the changes in place) that are occurring and how that affects people's emotional bonds to place. Intuitively vanishing places are locations that have been lost or humanity is at risk of losing. This is a new and emerging research theme within environmental psychology, but geographers are well positioned to continue contributing to this line of research given their proclivity for understanding human-environmental interactions and the tools at their disposal.

Based on ideas inspired by Stedman *et al.* (2014), I advance four categories of vanishing places. *Destruction* – These are places destroyed by a rapid, catastrophic event. The emphasis is placed on locations that are lost very quickly. The San Francisco earthquake and fire of 1906, Hurricane Katrina in 2005, and Super Storm Sandy in 2012 are all prime examples of vanishing places. On the night of May 4, 2007 an EF5 tornado destroyed 95 percent of the small, Great Plains' town of Greensburg, Kansas. With 23 percent of the town's population comprised of retired-aged people, this tragic event provided an opportunity to better understand how retirees develop strong feelings of attachment to place. Smith and Cartlidge (2011) found that key landmarks were incredibly important to retired residents because it helps them navigate through town, and age-specific businesses were vital to their connections to the community.

Depletion – These are places where there is a gradual loss of a natural resource or population; the emphasis is on slow change. From an environmental perspective, water depletion in the High Plains Aquifer and the retreating glaciers in Glacier National Park are two examples. Ben Marsh (1987) examined people's attachment to the coal mining districts of Appalachia and despite the depletion of coal in the region and the loss of viable employment opportunities, people refused to leave. Smith and McAlister (2015) explore the attachment that residents of the Great Plains have for their local county seat of government in the face of protracted population decline. A particularly timely topic ripe for intense scrutiny centers on how people respond to climate change (e.g., sea level rise). Preliminary research points to two responses including *in situ* adaption and environmental migration (see: Koubi *et al.* 2016).

Encroachment – These are places where outside cultural practices threaten the traditional character of the place. Ethnically homogenous neighborhoods in East Los Angeles is an example. Through the process of invasion and succession neighborhoods once dominated by African Americans have tipped and become largely Latino barrios. The same process is occurring again as Asian origin populations encroach upon the Latinos (see: Cheng 2013).

Restriction – These are places where certain activities are preferred or given preference over other activities. A long list of cultural geographers on both sides of the Atlantic including James and Nancy Duncan (2004); David Ley and Heather Smith (2000); Doreen Massey (1994); Gillian Rose (1990); and David Sibley (1995) have delved deep into this line of thinking.

In the final section of this book Geoffrey Buckley (Chapter 12) and Rex (RJ) Rowley (Chapter 13) provide two examples of vanishing places. Buckley explores how the Fountainbridge neighborhood in Edinburgh, Scotland is resisting change accompanied by growing tourism and seeking to hold on and honor its industrial past. Then, RJ Rowley draws upon his fieldwork in the fishing village of Kesennuma, Japan to explain how residents' connection to the sea is a double-edged sword. The 2011 earthquake and tsunami swept away most of what they knew in life, but their deep emotional ties to home gives them hope for the future. In the epilogue Paul Adams synthesizes the information shared and identifies common themes that transcend chapters and reinforce geography's contribution to place attachment research.

By placing emphasis on the PLACE component in place attachment research, we have an opportunity to make significant contributions to society. Not only will we better understand our individual identity and where we belong within society, but we can also foster a richer sense of community that enables people to develop emotionally healthy lives. City planners can design communities that are more user-friendly and accommodating to people of all ages and backgrounds. Place attachment research will also help us enact effective urban redevelopment and social housing projects. The possible applications for solid place attachment research are limited only by our imagination.

Geographers have the tools needed to make significant contributions to place attachment. We have an innate interest in understanding places, we seek to understand how people interact with places, and our discipline has always been one to approach scholarship from a holistic approach. Instead of approaching place attachment from the perspective of three distinct components (person, place, or process), geographers can help scholars in other disciplines see that a holistic approach might be more fruitful. This would put PLACE back in place attachment research because we would be seeking to understand how people, places, and processes all work together.

References

Agnew, J. A. 1987. *Place and Politics: The Geographical Mediation of State and Society.* Boston: Allen and Unwin.

Altman, I. 1975. *The Environment and Social Behavior.* Monterey, CA: Brooks & Cole.

Altman, I. and S. M. Low. 1992. *Place Attachment.* New York: Plenum Press.

Barcus, H. R. and S. D. Brunn. 2009. Towards a Typology of Mobility and Place Attachment in Rural America. *Journal of Appalachian Studies* 15: 26–48.

Blake, K. S. 1999. Peaks of Identity in Colorado's San Juan Mountains. *Journal of Cultural Geography* 18: 29–55.

Blake, K. S. 2008. Imagining Heaven and Earth at Mount of the Holy Cross, Colorado. *Journal of Cultural Geography* 25: 1–30.

Blake, K. S. 2010. Colorado Fourteeners and the Nature of Place Attachment. *Geographical Review* 92: 155–179.

Casey, E. S. 2001. Between Geography and Philosophy: What Does It Mean to Be in the Place-World? *Annals of the Association of American Geographers* 91: 683–693.

Chamlee-Wright, E. and V. H. Storr. 2009. "There's No Place Like New Orleans": Sense of Place and Community Recovery in the Ninth Ward after Hurricane Katrina. *Journal of Urban Affairs* 31: 615–634.

Cheng, W. 2013. *The Changs Next Door to the Díazes: Remapping Race in Suburban California.* Minneapolis: University of Minnesota Press.

Chow, K. and M. Healy. 2008. Place Attachment and Place Identity: First Year Undergraduates Making the Transition from Home to University. *Journal of Environmental Psychology* 28: 362–372.

Cresswell, T. 2004. *Place: A Short Introduction.* Oxford: Blackwell.

Devine-Wright, P. 2009. Rethinking NIMBYism: The Role of Place Attachment and Place Identity in Explaining Place-Protective Action. *Journal of Community and Applied Social Psychology* 19: 426–441.

de Wit, C. W. 2013. Interviewing for Sense of Place. *Journal of Cultural Geography* 30: 120–144.

Donohoe, J. 2014. *Remembering Places: A Phenomenological Study of the Relationship between Memory and Place.* Lanham, MD: Rowman & Littlefield Publishers.

Duncan, J. S. and N. Duncan. 2004. *Landscapes of Privilege: The Politics of the Aesthetic in an American Suburb.* New York: Routledge.

Entrikin, J. N. 2001. Hiding Places. *Annals of the Association of American Geographers* 91: 694–697.

Fauntleroy, G. 1997. La Querencia: The Place of Your Heart's Desire. *New Mexico Magazine* 75: 22–30.

Foote, K. E. 1997. *Shadowed Ground: America's Landscapes of Violence and Tragedy.* Austin: University of Texas Press.

Francaviglia, R. V. 2003. *Believing in Place: A Spiritual Geography of the Great Basin.* Reno: University of Nevada Press.

Fried, M. 1963. Grieving for a Lost Home. In *The Urban Condition: People and Policy in the Metropolis*, edited by L. J. Duhl, 151–171. New York: Basic Books.

Gillen, J. 2014. Tourism and Nation Building at the War Remnants Museum in Ho Chi Minh City, Vietnam. *Annals of the Association of American Geographers* 104: 1307–1321.

Hansen, S. 2008. Who Are "We"? An Important Question for Geography's Future. *Annals of the Association of American Geographers* 94: 715–722.

Hays, R. 1998. Sense of Place in Developmental Context. *Journal of Environmental Psychology* 18: 5–29.

Hernández, B., M. C. Hidalgo, and C. Ruiz. 2014. Theoretical and Methodological Aspects of Research on Place Attachment. In *Place Attachment: Advances in Theory, Methods, and Applications*, edited by L. Manzo and P. Devine-Wright, 125–137. London: Routledge.

Hernández, B., M. C. Hidalgo, M. E. Salazar-Laplace, and S. Hess. 2007. Place Attachment and Place Identity in Natives and Non-Natives. *Journal of Environmental Psychology* 27: 310–319.

Hoelscher, S. and D. H. Alderman. 2004. Memory and Place: Geographies of a Critical Relationship. *Social and Cultural Geography* 5: 347–355.

Howell, S. C. 1983. The Meaning of Place in Old Age. In *Aging and Milieu: Environmental Perspectives on Growing Old*, edited by G. D. Rowles and R. J. Ohta, 97–107. New York: Academic Press.

Jackson, J. B. 1994. *A Sense of Place, a Sense of Time.* New Haven, CT: Yale University Press.

Jorgensen, B. S. and R. C. Stedman. 2006. A Comparative Analysis of Predictors of Sense of Place Dimensions: Attachment to, Dependence on, and Identification with Lakeshore Properties. *Journal of Environmental Management* 79: 316–327.

Koubi, V., S. Stoll, and G. Spilker. 2016. Perceptions of Environmental Change and Migration Decisions. *Climate Change* 138: 439–451.

Lewicka, M. 2005. Ways to Make People Active: The Role of Place Attachment, Cultural Capital, and Neighborhood Ties. *Journal of Environmental Psychology* 25: 381–395.

Lewicka, M. 2011. Place Attachment: How Far Have We Come in the Last 40 Years? *Journal of Environmental Psychology* 31: 207–230.

Lewis, P. 1979. Defining a Sense of Place. *Southern Quarterly* 17: 24–46.

Lewis, P. 1985. Beyond Description. *Annals of the Association of American Geographers* 75: 465–477.

Ley, D. and H. Smith. 2000. Relations between Deprivation and Immigrant Groups in Large Canadian Cities. *Urban Studies* 37: 37–62.

Low, S. M. and I. Altman. 1992. Place Attachment: A Conceptual Inquiry. In *Place Attachment*, edited by I. Altman and S. M. Low, 1–12. New York: Plenum Press.

Lowenthal, D. 2015. *The Past Is a Foreign Country Revisited.* Cambridge: Cambridge University Press.

Manzo, L. C. and P. Devine-Wright. 2014. *Place Attachment: Advances in Theory, Methods, and Applications*. London: Routledge.

Marcus, C. C. 1995. *House as a Mirror of Self: Exploring the Deeper Meaning of Home*. Berwick, ME: Nicolas-Hays, Inc.

Marsh, B. 1987. Continuity and Decline in the Anthracite Towns of Pennsylvania. *Annals of the Association of American Geographers* 77: 337–352.

Massey, D. 1994. *Space, Place, and Gender*. Minneapolis: University of Minnesota Press.

Massey, D. 2005. *For Space*. London: Sage Publications.

Mattila, A. S. 2001. Emotional Bonding and Restaurant Loyalty. *Cornell Hotel and Restaurant Administration Quarterly* 42: 73–79.

McAndrew, F. T. 1998. The Measurement of "Rootedness" and the Prediction of Attachment to Home-Towns in College Students. *Journal of Environmental Psychology* 18: 409–417.

Meade, M. S. and M. Emch. 2000. *Medical Geography*, 3rd ed. New York: Guilford Press.

Mihaylov, N. and D. D. Perkins. 2014. Community Place Attachment and Its Role in Social Capital Development. In *Place Attachment: Advances in Theory, Methods, and Applications*, edited by L. Manzo and P. Devine-Wright, 61–74. London: Routledge.

Milligan, M. J. 1998. Interactional Past and Potential: The Social Construction of Place Attachment. *Symbolic Interaction* 21: 1–33.

Murphy, A. 2014. Geography's Crosscutting Themes: Golden Anniversary Reflections on "The Four Traditions of Geography." *Journal of Geography* 113: 181–188.

Nasar, J. L. and D. A. Julian. 1995. The Psychological Sense of Community in the Neighborhood. *Journal of the American Planning Association* 61: 178–184.

Nostrand, R. L. 1992. *The Hispano Homeland*. Norman: University of Oklahoma Press.

Nostrand, R. L. and L. E. Estaville, Jr. 1993. Introduction: The Homeland Concept. *Journal of Cultural Geography* 13: 1–4.

Nostrand, R. L. and L. E. Estaville, Jr. 2001. *Homelands: A Geography of Culture and Place across America*. Baltimore: Johns Hopkins University Press.

Oldenburg, R. 1989. *The Great Good Place: Cafes, Coffee Shops, Bookstores, Bars, Hair Salons, and Other Hangouts at the Heart of a Community*. New York: Marlowe and Company.

Pattison, W. 1964. The Four Traditions of Geography. *Journal of Geography* 63: 211–216.

Perkins, D. D. and D. A. Long. 2002. Neighborhood Sense of Community and Social Capital: A Multi-Level Analysis. In *Psychological Sense of Community: Research, Applications, and Implications*, edited by A. Fisher, C. Sonn, and B. Bishop, 291–318. New York: Plenum.

Ponzetti, Jr., J. J. 2003. Growing Old in Rural Communities: A Visual Methodology for Studying Place Attachment. *Journal of Rural Community Psychology* 6: 1–11.

Post, C. 2008. Modifying Sense of Place in a Federal Company Town: Sunflower Village, Kansas, 1942 to 1959. *Journal of Cultural Geography* 25: 137–159.

Pretty, G. H., H. M. Chipuer, and P. Bramston. 2003. Sense of Place Amongst Adolescents and Adults in Two Rural Australian Towns: The Discriminating Features of Place Attachment, Sense of Community, and Place Dependence in Relation to Place Identity. *Journal of Environmental Psychology* 23: 273–287.

Proshansky, H. M. 1978. The City and Self-Identity. *Environment and Behavior* 10: 147–169.

Proshansky, H. M., A. K. Fabian, and R. Kaminoff. 1983. Place-Identity. *Journal of Environmental Psychology* 3: 57–83.

Raymond, C. M., G. Brown, and D. Weber. 2010. The Measurement of Place Attachment: Personal, Community, and Environmental Connections. *Journal of Environmental Psychology* 30: 422–434.

Relph, E. 1976. *Place and Placelessness*. London: Pion Limited.

Rivlin, L. G. 1982. Group Membership and Place Meanings in an Urban Neighborhood. *Journal of Social Issues* 38: 75–93.

Romero, L. 2001. La Nueva Resolana. *New Mexico Magazine* 79: 26–31.

Rose, G. 1990. The Struggle for Political Democracy: Emancipation, Gender, and Geography. *Environment and Planning D: Society and Space* 8: 395–408.

Rowles, G. D. 1990. Place Attachment Among the Small Town Elderly. *Journal of Rural Community Psychology* 11: 103–120.

Rubenstein, R. L. and P. A. Parmelee. 1992. Attachment to Place and the Representation of the Life Course by the Elderly. In *Place Attachment*, edited by I. Altman and S. M. Low, 139–163. New York: Plenum Press.

Sack, R. D. 1988. The Consumer's World: Place as Context. *Annals of the Association of American Geographers* 78: 642–664.

Scannell, L. and R. Gifford. 2010. Defining Place Attachment: A Tripartite Organizing Framework. *Journal of Environmental Psychology* 30: 1–10.

Scannell, L. and R. Gifford. 2014. Comparing the Theories of Interpersonal and Place Attachment. In *Place Attachment: Advances in Theory, Methods, and Applications*, edited by L. Manzo and P. Devine-Wright, 23–36. London: Routledge.

Schnell, S. M. 2003. Creating Narratives of Place and Identity in "Little Sweden, U.S.A." *Geographical Review* 93: 1–29.

Seamon, D. 2008. Place, Placelessness, Insideness, and Outsideness in John Sayles' Sunshine State. *Aether* 3: 1–19.

Seamon, D. 2014. Place Attachment and Phenomenology: The Synergistic Dynamism of Place. In *Place Attachment: Advances in Theory, Methods, and Applications*, edited by L. Manzo and P. Devine-Wright, 11–22. London: Routledge.

Seamon, D. and J. Sowers. 2008. Place and Placelessness, Edward Relph. In *Key Texts in Human Geography*, edited by P. Hubbard, R. Kitchen, and G. Vallentine, 43–51. London: Sage Publications.

Shortridge, J. R. 1989. *The Middle West: Its Meaning in American Culture*. Lawrence: University of Kansas Press.

Shumaker, S. A. and R. B. Taylor. 1983. Toward a Clarification of People-Place Relationships: A Model of Attachment to Place. In *Environmental Psychology: Directions and Perspectives*, edited by N. R. Feimer and E. S. Geller, 219–256. New York: Praeger.

Sibley, D. 1995. *Geographies of Exclusion: Society and Difference in the West*. London: Routledge.

Sifferlin, A. 2016. The Healing Power of Nature. *TIME* 188: 24.

Smith, J. S. 2004. The Plaza in Las Vegas, New Mexico: A Community Gathering Place. In *Hispanic Spaces, Latino Places: Community and Cultural Diversity in Contemporary America*, edited by D. D. Arreola, 39–54. Austin: University of Texas Press.

Smith, J. S. and M. Cartlidge. 2011. Place Attachment among Retirees in Greensburg, Kansas. *Geographical Review* 101: 536–555.

Smith, J. S. and J. M. McAlister. 2015. Understanding Place Attachment to the County in the American Great Plains. *Geographical Review* 105: 178–198.

Smith, J. S. and B. N. White. 2004. Detached from Their Homeland: The Latter-Day Saints of Chihuahua, Mexico. *Journal of Cultural Geography* 21: 57–76.

Stedman, R. C., B. L. Amsden, T. M. Beckley, and K. G. Tidball. 2014. Photo-Based Methods for Understanding Place Meaning as Foundations of Attachment. In *Place Attachment: Advances in Theory, Methods, and Applications*, edited by L. Manzo and P. Devine-Wright, 112–124. London: Routledge.

Tuan, Y.-F. 1974. *Topophilia: A Study of Environmental Perception, Attitudes, and Values*. Englewood Cliffs, NJ: Prentice-Hall, Inc.

Tuan, Y.-F. 1976. Geopiety: A Theme in Man's Attachment to Nature and to Place. In *Geographies of the Mind: Essays in Historical Geosophy in Honor of John Kirtland Wright*, edited by D. Lowenthal and M. J. Bowden, 11–39. New York: Oxford University Press.

Tuan, Y.-F. 1977. *Space and Place: The Perspective of Experience*. Minneapolis: University of Minnesota Press.

Tuan, Y.-F. 1980. Rootedness versus Sense of Place. *Landscape* 24: 3–8.

van der Graaf, P. 2008. *Out of Place? Emotional Ties to the Neighborhood in Urban Renewal in the Netherlands and the United Kingdom*. Vossiuspers: University of Amsterdam Press.

Weil, S. 1952. *The Need for Roots*. Boston: Beacon Press.

Williams, D. R. 2014. "Beyond the Commodity Metaphor," Revisited: Some Methodological Reflections on Place Attachment Research. In *Place Attachment: Advances in Theory, Methods, and Applications*, edited by L. Manzo and P. Devine-Wright, 89–99. London: Routledge.

Wishart, D. J. 2013. *The Last Days of the Rainbelt*. Lincoln: University of Nebraska Press.

Wright, J. K. 1947. Terrae Incognitae: The Place of the Imagination in Geography. *Annals of the Association of American Geographers* 37: 1–15.

Wyckoff, W. and L. M. Dilsaver. 1995. *The Mountainous West: Explorations in Historical Geography*. Lincoln: University of Nebraska Press.

Young, T. 2001. Place Matters. *Annals of the Association of American Geographers* 91: 681–682.

Part I

Secure places

1 Influence of memory on post-resettlement place attachment

Michael Strong

Despite broad attention to place attachment by environmental psychologists over the past 40 years (Lewicka 2011) and the emergence of useful ways to organize studies of place attachment (see: Scannell and Gifford 2010), the interrogation and interpretation of the *place* dimension within place attachment has been woefully under-theorized. To be fair, environmental psychologists acknowledge the contributions of select humanist geographers, especially Relph (1976) and Tuan (1974), but they often promote a study of psychological affect, cognition, and behavior within place at the expense of studying the nature of the places to which their subjects become attached. Despite this, the intersection of psychology and geography can be fruitful in the pursuit of better understanding people's emotional ties to place.

Individuals are most likely to develop strong emotional ties to a place where they feel safe and secure (Scannell and Gifford 2010). Home is a quintessential example. As people are confronted with a new location, they evaluate the new place by making unconscious comparisons to their home. Should a community find itself forced to relocate to a new place, a case explored in this chapter, members of the community seek opportunities to transform the new place into something that resembles the place they lost. This suggests that place attachment cannot form until an individual sees the place as home. Interestingly, this does not have to be a physical transformation.

In Mozambique's Tete Province, residents have been forced to relocate from ancestral homelands near the Moatize River because the Mozambican government granted permission to an international coal mining company to exploit local deposits. This forced relocation has offered an opportunity to examine how resettled residents have responded to life in a new location. Through storytelling and political action, many residents have found ways to bring images of their old home to the new location. This chapter seeks to explain how memories of home can offer a sense of safety and security in an environment that differs substantially from the place lost. By doing so, I call attention to the power of the place. In other words, the relationship one has with past places can influence how one interacts with present and future places. In this chapter, I first situate place and memory within the literature on toponymy and resettlement before describing the study's methods. After presenting the results, I conclude with some thoughts on the power of place (operating through memories) to influence the outcomes of resettlement.

Literature

Place, memory, and place attachment

Geographers describe place as bounded space that is meaningful to its user (Relph 1976; Tuan 1974), but place is more than its location; it is also the product of an emotional connection to the landscape and environment (Cresswell 2004; Relph 1976; Trigg 2012), such that place transcends its mere spatial location to connect humans to space as "the center of felt value" (Tuan 1974, 4). Given the importance that a place plays in the shaping of a person's identity, it is easily discernible how a disruption to a person's relationship to place can be traumatic.

The power of place permeates popular culture. Nowhere is this more evident than in the 1939 classic film, *The Wizard of Oz*. Despite Kansas as a black-and-white landscape of desiccated, tornado-prone flatness where an evil neighbor wants to kill her dog, Dorothy learns – through her adventures in Oz's Technicolor dream-world – that *there is no place like home*. Each new encounter with talking trees and winged monkeys, with dancing scarecrows and melting witches, presents Dorothy with an opportunity to compare her home in Kansas to the many diverse but quite different places of Oz. Dorothy learns this lesson alongside all those who view the film such that the message resonates not just with the protagonist but also with the audience.

Dorothy's experience is not unfamiliar. In his *Poetics of Space*, Bachelard (1957/1994) posits that places follow a person from location-to-location, predisposing future encounters with similar locations to be unconscious comparisons of prior experience(s) within the places of one's past. Encountering a place for the first time is both novel and not. The various aspects of the place (e.g., doors, windows, trees, sidewalks, train tracks, etc.) are unique yet also notably familiar (Heidegger 1996). A really comfortable chair is only comfortable when compared to all the uncomfortable chairs of one's past.

In a way, places exhibit a characteristic resembling Soja's (1989) spatiality. Just as people interacting with space can change both into something different, the interaction of place with memory can result in the outcome of place memory (Casey 1987). Places, as meaningful space, hold special significance for individuals, but places also hold individual and collective memory (Casey 2004). Particularly poignant encounters with place become embedded within the spaces of the brain's neural networks available for recall when the appropriate connotative signals draw them from the recesses of memory into the here-and-now of experience. Imagine a young graduate arriving late for an important job interview because a commuter train was undergoing repairs at a busy station. If she believes that the repairs cost her the job, it is possible that every future encounter with that same train station will prompt a recall of that negative experience.

Sociologist Maurice Halbwachs (1980) empowered place as the repository of collective memory, stating that in places "the collective thought of the group of believers has the best chance of immobilizing itself and enduring . . . [by] sealing itself within their confines and molding its character to theirs" (presented in Casey 2004, 44). Monuments, memorial plazas, landmarks, placards, and other public structures collect the past and influence the development of shared memories – even for events participants have never personally experienced. Some of these repositories of place memory (e.g., monuments) are tangible while still others have only a fleeting pseudo-tangibility. Place names comprise this latter category of place memory.

Memories of the places where one lived in the past provide opportunity to "situate [oneself] in an idyllic social and physical landscape" (McMichael 2003, 95). Thus, it becomes possible to compare and contrast present places with those of the past through the nostalgic lens of memory. Nostalgia involves constructing memories by filtering out negative experiences while honing in on positive ones (Davis 1973). In a way, nostalgia promotes a longing for the past because the present is so disconcerting (Dann 1994). Feeling nostalgic, though, can evince positive change by using the positivity of the past as a means to alter the environment in some way as to construct a better present and future (Spitzer 1999).

One such way is to rename an undesirable place with a desirable name invoking positive place memories. However, naming decisions are not apolitical. Toponymic approaches to geographic study stress the embedded power struggles and social relations present in naming decisions (Berg and Kearns 1996; Berg and Vuolteenhaho 2009). Social change and political revolution often result in pushes to rename places in a way that erases their troubled past or commemorates individual and collective identity (Light 2004; Guyot and Seethal 2007). Reuben Rose-Redwood *et al.* (2010) call specifically for greater attention to how places are assigned names to ensure the name represents the social identity and collective history of the place's inhabitants. This is especially important in Mozambique where, as João Baptista (2010) explains, it is very difficult to separate the story of the place from the story of the group living there.

Thus, it makes perfect sense as to why a resettled population would seek to rename the place to which it has resettled. Names are a form of pseudo-tangible collective memory imbued with the social relations, cultural identity, and history of the society that inhabits the place associated with the name. But, what happens when a group's new home does not reflect the same social and physical features of the place associated with the name? The memories of the past place can conflict with the realities of the present place to hinder the establishment of place attachment for some members of the resettled community. This is worthy of our attention. Individuals without the ability to form an emotional attachment to place experience anxiety (Casey 2009) and distress (Fried 1963), not to mention it can lead to an entire generation of placeless individuals (Strong 2016). Unfortunately, in Mozambique, the potential for placeless populations to arise following resettlement is quite common given that state's propensity to employ resettlement under a wide variety of situations.

Resettlement in Mozambique

Development-induced displacement and resettlement (DIDR) directly impacts more than 15 million people each year worldwide (Bugalski and Pred 2013). In early 2015, Jim Yong Kim, President of the World Bank, described DIDR as an inescapable reality if countries are to meet demand for infrastructure and predicted that the number of displaced individuals will continue to rise (Donnan 2015). This is troubling news. Resettled populations frequently face a number of risks that can lead to even more impoverishing conditions in the post-resettlement community than had previously existed (Cernea 1997). This is a significant burden for developing countries and funding partners to bear, notwithstanding the potential impact on the lives of so many already impoverished people.

Resettlement has a long history in Mozambique. Prior to independence, the Portuguese government embarked upon a villagization program to solidify control over rural communities (Borges Coelho 1998). During the civil war, the government concentrated rural residents in villages as a means to ensure their protection (Lubkemann 2008). Today, resettlement continues to be a routine practice of the government: to mitigate risk following flooding (Stal 2011; Artur and Hilhorst 2014), to promote conservation efforts (Milgroom and Spierenburg 2008), and to permit mining companies to exercise mining concessions (Kirshner and Power 2015; Lillywhite *et al.* 2015). Although the Resettlement Decree provides for compensation to address losses (Republic of Mozambique 2012), the Mozambican government has failed to adequately compensate resettled populations for the invisible losses they have incurred (Witter and Satterfield 2014). How does one put a price on the personal relationships people develop to specific places and the inability to (re)-form these connections in the post-resettlement site?

Mozambique remains one of the poorest countries in the world (United Nations 2014; World Bank 2015). Despite this, Mozambique has experienced impressive economic growth over the last decade (African Economic Outlook [AEO] 2015). Since 2008, multinational corporations have invested billions in mining infrastructure in the Moatize District, leading some to refer to the region as Mozambique's *El Dorado* (Mosca and Selemane 2011). In Mozambique, local communities can petition the government for the exclusive right to land and resources, but under the National Land Law, the federal government has the authority to grant land use concessions to corporations as well. In the Moatize District, the national government has allocated nearly 75 percent of the region's land to multinational corporations for resource extraction (Human Rights Watch 2013). To exercise mining concessions, companies have already resettled several villages, including the Bairro Chipanga resettlement (Lillywhite *et al.* 2015).

Case study site

Between 2009 and 2010, the Brazilian multinational corporation, Vale, resettled 289 households from four peri-urban villages in the Moatize District to Bairro Chipanga (Figure 1.1). Prior to resettlement, family members typically worked as stone masons, brick makers, mechanics, carpenters, electricians, and small-scale vendors (Selemane 2010). Most of Bairro Chipanga's new families originated from Chipanga, a village of the same name approximately 9 km southeast of central Moatize (Pedro 2011). Most of the resettled residents are members of the

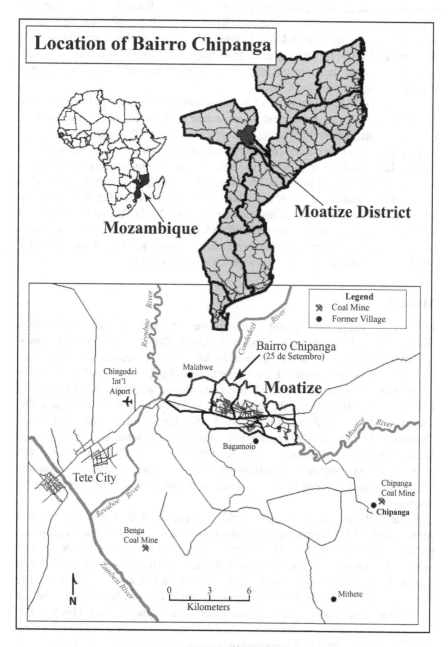

Figure 1.1 Location of Bairro Chipanga and former villages.

Cartography by author

Nyungue ethnic group, a patrilineal population with traditional marriage practices. Wives in most polygamous households lived near their husband but on separate land plots with their own children. Households ranged in size from four to eight persons (Pedro 2011). Chipanga residents had access to a health clinic, primary school, recreation field, various churches, a marketplace, and Belo Horizonte (a night club popular with local youth).

Though resettlement to Bairro Chipanga was optional (the government also offered residents a cash payment or assistance with finding a house elsewhere) (Human Rights Watch 2013), those residents who relocated to Bairro Chipanga received a newly constructed concrete house on a poured foundation capped with a zinc-plated roof (Gerety 2013). The houses were accompanied by an open-air kitchen, bathhouse and toilet facility, electricity, and piped water; larger families obtained a second house if they had adult children or extended family members in their household at the time of resettlement (Gerety 2013). At first compensation for the loss of farming plots was not offered, but in January 2014 Vale agreed to provide compensation because the former farming plots were too distant for many residents to continue using the same plots (AllAfrica 2014).

Methods

Data for this chapter originated primarily from 75 interviews collected in 2015 from residents living in Bairro Chipanga. Using 2014 satellite imagery down-loaded from Google, all dwellings built by Vale to house resettled villagers were mapped and assigned a number. Every fourth household was selected for inclusion in the study. Three trained research assistants interviewed the head of household in the preferred language of the respondent (either Portuguese or Nyungue). The research assistants collected a wide variety of information including household characteristics, perspectives on the pre- and post-settlement communities, and feelings of attachment to place. To protect the identity of the respondents, each respondent was assigned a numerical identification code. All data were retained in a secure location. Table 1.1 presents a demographic overview of the households interviewed.

To gauge the importance of place, residents were asked open-ended questions in semi-structured interviews. My analysis of the data followed guidelines proposed by Miles and Huberman (1994). Prior to reading the interviews, I developed an initial set of codes related to the general theme of each open-ended question that would characterize how the respondent related to the place they lived (e.g., life in former villages, life in Bairro Chipanga, risks in Bairro Chipanga that did not exist before resettlement, preferred attributes of both places, etc.). This was fol-lowed by a careful and iterative reading of the interview responses to look at how these themes emerged in the context of the question asked. During this process, additional codes emerged that supplemented, and sometimes replaced, the codes developed *a priori* to the analysis. I used the text that emerged from the coding to understand how memory of the former Chipanga influenced the formation of place attachment to Bairro Chipanga.

Table 1.1 Household characteristics of respondents (*n* = 75)

	#	%	Mean	SD	Min	Max
Age			43.9	12.0	19	73
Education (years of schooling)			5.6	3.8	0	12
Employment (= formal)	26	34.7				
Gender (= female)	29	38.7				
Household size			7.3	2.9	1	16
Marital status (= married)	65	86.7				
Number of school-age children (6–16)			3.5	2.7	0	12
Resettled from:						
Bagamoio	6	8.0				
Chipanga	65	86.7				
Malabwe	1	1.3				
Mithete	3	4.0				

Author's tabulations

The two Chipangas

To say the residents of Bairro Chipanga expressed fond memories of the former Chipanga would be a tremendous understatement. Residents described Chipanga as a place where they "lived well" and had a "good life" or "good relationships." Residents spoke of ample cropland and the productivity of their *machambas* (farming plots). They described the easy access to water via the river and public fountain, and explained how the adjacent bush provided plentiful resources. They mentioned the helpfulness of neighbors and the way that everyone "lived as a family." Chipanga was consistently identified as a place where residents enjoyed positive social interactions and a fruitful physical landscape. Resources were plentiful and neighbors did whatever they could to help one another. In Chipanga, opportunities abounded. For example, if a resident did not have a job, they could find temporary work making bricks and "transporting those bricks on their heads." Most importantly, residents did not need cash to access basic necessities. Firewood, water, and farmland were all free and readily available. When asked about life in Bairro Chipanga, the reverse was true. Many residents believed the resettlement was "bad for everyone in the village," and words commonly used to describe life in Bairro Chipanga include "challenging" and "insufferable."

Given how negatively residents viewed life in Bairro Chipanga, I encountered a curious source of incongruence within the community. Official paperwork on the resettlement process as well as previous research reports consistently referred to the site as "Unidade 6 of Bairro 25 de Setembro" (Pedro 2011; Gerety 2013; Human Rights Watch 2013). Official signage in the area, as well as information published by Vale, also reflects this official place name (Figure 1.2). When I first visited the settlement in January 2015, I was welcomed into the home of the local leader who spoke of Unidade 6. However, the place name consistently used by local residents, key informants, and even my research assistants when I returned in May 2015 was Bairro Chipanga. My lead research assistant, who also worked

Figure 1.2 Center of the community, Unidade 6 (Bairro Chipanga).
Photo by author, 2015

as a local primary school teacher, told me that a lot had changed in the five months between my two visits. The residents of this "insufferable" settlement had finally been successful in their petition to the provincial government to change the place name and enact local self-governance.

During the initial planning of the resettlement process, the government had named the post-resettlement site Unidade 6 of Bairro 25 de Setembro. In Mozambique, many urban areas (including Moatize) are divided into *unidades* (wards) where a local, traditional leader provides a separate governing structure from the district-level government. Following the conclusion of the resettlement process, traditional leaders from the former village of Chipanga (which comprised approximately 90 percent of the resettled population) asked for political independence for the resettlement site (i.e., the designation of a new and distinct bairro within Moatize). Their argument was simple. They reasoned it was logical that a community resettled from one place to another should have the right to maintain its governance status post-resettlement.

Speaking about motivation for the name change, the woman selected as the leader of Bairro Chipanga, told me: "We refused to be considered as a sub-area [to Bairro 25 de Setembro]. Why were they incorporating us into another village? Do [that village's leaders] know our ancestor's spirits?" This lack of knowledge of the people of Chipanga, who by this point included those residents resettled from the other villages, also manifested in the way the leadership in Bairro 25 de Setembro responded to the concerns of residents in Bairro Chipanga. As one older man said, "When we reported our own worries to the chief of 25 de Setembro, he ignored us because he wasn't from the former Chipanga village." Given that residents of the resettled community felt like the leadership of Bairro 25 de Setembro could not, and did not, relate to the resettled group, it is no surprise they sought political independence.

Many residents in Bairro Chipanga spoke positively of the decision to make Unidade 6 into a new bairro of Moatize. When asked how the bairro came to have the name of the former village, nearly half the people surveyed (including those who had resettled from one of the three other villages) specifically stated they were happy to see the new bairro named in honor of their former village, even though many were unable to describe exactly how the change had happened. For these residents, choosing the Chipanga name was important to them.

As reported in Table 1.2, not everyone shared the perspective that Chipanga was an appropriate name for the new bairro. A few residents disagreed with the choice of Bairro Chipanga. One respondent who had been a traditional leader before the resettlement had suggested an alternative name to the one selected.

> Yes, [I suggested] Bairro Chipanga Nova [meaning "New Chipanga"]. The reason is that it is a new Chipanga, a Chipanga with a different vision, reality, and appearance. Why would we give it the same name? Chipanga is the village we left. Did they give it this name just because some people came from Chipanga? Are all the people from Chipanga village? To me, this idea wasn't good.

Importantly, what made Chipanga a poor selection for the name of the newly independent bairro was the post-resettlement site's incompatibility with the Chipanga name. As a former village leader explained, "When we call it Chipanga, it sounds as if everything in Chipanga is available here. I can't go along with this idea." Other residents echoed this sentiment; the name insinuated an inappropriate comparison to the former village. One young man said, jokingly, "It is nonsense to just name it Chipanga while there is nothing good in here." A much older resident

Table 1.2 Perspectives on renaming the resettlement site

Theme from Interview (n = 75)	N	%
Changing name to Chipanga was a good idea.	31	41.3
Chipanga name is important to me.	18	24.0
Chipanga was not my choice for name.	7	9.3
I am aware of the name change.	59	78.7
I like the Chipanga name.	34	45.3
New name changes nothing about the bairro.	14	18.7
New name is not an official change.	8	10.7
The government decided to change name.	13	17.3

Author's tabulations

echoed his sentiment: "Nothing is available here. They [the government] have only copied the name of our motherland." One skeptical young man suggested that the government only approved the name to "comfort the villagers [even though] it will not help to minimize the problems. [He believed] they can change the name, but if they don't change their attitudes, nothing will help."

The incompatibility of Bairro Chipanga with the Chipanga name rests in the place characteristics of the post-resettlement site. During the semi-structured interviews, residents were asked to describe risks they faced living in Bairro Chipanga that they had not faced before the resettlement. Notably, "access to water" and "lack of food security" featured prominently in residents' responses. Both represent the loss of access to resources that were free and plentiful in the former Chipanga. In Bairro Chipanga, residents have to pay for water and buy food to survive. The Revuboe River abuts the northern boundary of the bairro, but it is too far to travel for widespread water access; in the former Chipanga, the Moatize River ran through the village, and everyone had regular access to government-provided fountains. The housing plots in Bairro Chipanga are sufficiently large enough to plant small vegetable gardens, but without regular access to water, productivity is limited unless residents pay for water from their tap. In the former Chipanga, machambas abutted the Moatize River and were only a short walk from home with plentiful access to water.

Finally, as an urban environment, Bairro Chipanga does not have any adjacent bush areas where residents can collect needed resources (e.g., firewood); instead, they must buy these resources from the local market or from street vendors. In contrast, the former Chipanga was located at the edge of the bush and firewood was plentiful. The local environment also supported a variety of small-scale businesses that supplemented household income from crop sales. As an older widowed woman told us, "We collected sand to sell and went into the forest to collect firewood. We sold them and got some money. We had many businesses, such as producing and selling charcoal, or making bricks for sale." These enterprises are not possible in Bairro Chipanga.

Renaming the settlement: place attachment and memory

Among those residents against the selection of Chipanga as the bairro's new name, opposition was not a reflection of Chipanga as a suitable name but rather the suitability of the post-resettlement site to receive the Chipanga name. In the oft-quoted drama *Romeo and Juliet*, Shakespeare's Juliet laments that her beloved Romeo's surname is an enemy of her house: "What's in a name? That which we call a rose/By any other word would smell as sweet*"* (Act II, Scene II, lines 43–44, *The Riverside Shakespeare*, edited by G. Blakemore Evans). While this may be true for roses, the same cannot be said of the post-resettlement site selected by Vale in northwestern Moatize. Residents spoke quite strongly about the negative characteristics of Bairro Chipanga and how it differed from their former village of Chipanga. If there were really nothing positive about the place characteristics of the post-resettlement site, at least in comparison to the former Chipanga, why did the community exert so much effort to declare independence from Bairro 25 de Setembro and seek to rename Unidade 6 in honor of their former village?

Echoing Rose-Redwood *et al.*'s (2010) call for greater attention to the spatial politics of place name decisions, these results suggest there was good reason for the residents to select Bairro Chipanga as the new name for the post-resettlement site. Unidade 6 does not describe anything specific about the place or the people who inhabit it. There is no historical affiliation between Unidade 6 and Moatize except that the former only came into existence after the resettlement process concluded. In Mozambique, the history of a rural population is often inseparable from the history of the place that population inhabits (Baptista 2010). By naming the post-resettlement site Bairro Chipanga, the community imbued the post-resettlement site with a history and identity that predated its actual formation in 2010. The renaming also provided a means for the community to reclaim control over its future. The name was not just a commemoration of the past but also an indication that political independence had been all but granted. It signified the end of a shared struggle for post-resettlement identity and the start of a new chapter in a new, but familiar, place.

The need to forge an identity in the post-resettlement site was uncontested in Bairro Chipanga; however, what that identity should be was not. Many residents favored retaining the Chipanga identity, an unsurprising finding given Bachelard's (1957/1994) work on how places follow individuals through space and time. While Bachelard's writings are useful for understanding the population's desire to select Chipanga as the name for the post-resettlement site, I refer to Heidegger's (1996) work to understand the opposition. Heidegger writes that encounters with new spaces are (un)conscious comparisons to all the places of one's past. The same can be said for the residents of Bairro Chipanga. Lacking a physical environment reflecting the ways that residents interacted with the places of their past, the post-resettlement site exhibited little reason for residents to develop a relationship to the site, and it certainly was unlikely to become "home." This is a fact problema-tized by memories of the former Chipanga. Interacting with the spaces of Bairro Chipanga, residents were reminded of what they had lost in the resettlement: land to plant a machamba, free water and various bush resources (like firewood), and opportunities for small-scale business enterprises linked to the physical geography of their home.

Through the process of renaming the post-resettlement site, the residents not only imbued the place with a history and an identity, they also empowered it as a repository of their memories (Halbwachs 1980). Unfortunately, unlike the situa-tion described by McMichael (2003), the intersection of memory with Bairro Chi-panga completely failed to situate the population in an "idyllic social and physical landscape" (95). On the contrary, the renaming process engendered the growth of nostalgia for something that had been totally lost, and like Davis (1973) writes, probably never actually existed. Throughout the interviews, residents consistently spoke of the former Chipanga as the *El Dorado* others use to semi-facetiously describe contemporary Tete Province (Mosca and Selemane 2011; Kirshner and Power 2015). I visited the former Chipanga area and could not discern much dif-ference between the two places. The sunset was quite lovely, and the image of children playing in the Moatize River made for a great picture, but the road to the former Chipanga was no Shangri-La. Instead, what I observed was the use of nostalgia to selectively filter residents' place memory to construct a place that truly only existed in the collective memory of the community.

For the residents of Bairro Chipanga, it was not the name of the post-resettlement site that was really important but rather the memory connoted with the name. In the short time that Unidade 6 existed, it failed to provide a context in which the residents could find a *raison d'être*. That *raison* had existed in the former Chipanga, and so it was to this that the residents turned when they declared independence and proclaimed not Chipanga Nova (as was suggested by a former traditional leader), but Chipanga, a phoenix arisen from the ashes of the past into new life on the edge of Moatize. Of course, when it was just Unidade 6, it was a symbol of the resettlement. When it became Chipanga, it became a symbol of both the resettlement and of everything that the people had left behind.

This last point is very telling for the role that memory plays in the formation of place attachment. Despite what Shakespeare writes, there is power in a name. The residents of Bairro Chipanga carried their former home to the post-resettlement site through their individual and collective memories of the former Chipanga. Nostalgia for the former village prompted a movement to seek political independence within the governing structure of Moatize. As Spitzer (1999) writes, feeling nostalgic for past places can push individuals to create a better present and future. Through the process of renaming the bairro in honor of their former village, the residents of Bairro Chipanga imbued a placeless place with a physical and social geography it had previously lacked. They made it into a secure place. They made it home.

While Bairro Chipanga would never have the same attributes as the former Chipanga, at least it could have the former Chipanga's name: a name that evokes memories of happiness, the good life, and a good place. It is a name that evokes memories of home. These feelings are important if one is to develop a sense of safety and security, critical features of a home and necessary precursors to the formation of place attachment. Resettlement planners, governments, corporations, and other agents involved in resettlement planning could greatly enhance this process if they design a resettlement site to include features that remind the resettled community of its former home. In this way, they can call upon the power of place memories to ease resettled populations' transition from one place to another.

References

African Economic Outlook [AEO]. 2015. *Statistics Table 2: Real GDP Growth Rates, 2006–2016*. Immeuble Seine Saint Germain, France: OECD Development Centre. www.africaneconomicoutlook.org/en/statistics/. Last accessed 24 October 2015.

AllAfrica. 16 January 2014. Vale Agrees to Further Compensation in Moatize. *AllAfrica*. http://allafrica.com/stories/201401161592.html. Last accessed 21 October 2015.

Artur, L. and D. Hilhorst. 2014. Floods, Resettlement, and Land Access and Use in the Lower Zambezi, Mozambique. *Land Use Policy* 36: 361–368.

Bachelard, G. 1957/1994. *The Poetics of Space*. Boston: Beacon Press.

Baptista, J. 2010. Disturbing "Development": The Water Supply Conflict in Canhane, Mozambique. *Journal of Southern African Studies* 36: 169–188.

Berg, L. and R. Kearns. 1996. Naming as Norming: "Race", Gender, and the Identity Politics of Naming Places in Aotearoa/New Zealand. *Environment and Planning D* 14: 99–122.

Berg, L. and J. Vuolteenhaho (eds.). 2009. *Critical Toponymies: The Contested Politics of Place Naming*. Burlington: Ashgate.

Borges Coelho, J. 1998. State Resettlement Policies in Post-Colonial Rural Mozambique: The Impact of the Communal Village Programme on Tete Province, 1977–1982. *Journal of Southern African Studies* 24: 61–91.

Bugalski, N. and D. Pred. 2013. *Reforming the World Bank Policy on Involuntary Resettlement*. Washington, DC: Inclusive Development International.

Casey, E. 1987. *Remembering: A Phenomenological History*. Berkeley: University of California Press.

Casey, E. 2004. Public Memory in Place and Time. In *Public Memory*, edited by K. Phillips, 17–44. Tuscaloosa, AL: University of Alabama Press.

Casey, E. 2009. *Getting Back into Place: Toward a Renewed Understanding of the Place-World*, 2nd ed. Bloomington, IN: Indiana University Press.

Cernea, M. 1997. The Risks and Reconstruction Model for Resettling Displaced Populations. *World Development* 25: 1569–1587.

Cresswell, T. 2004. *Place: A Short Introduction*. Malden, MA: Blackwell Publishing.

Dann, G. 1994. Tourism: The Nostalgia Industry of the Future. In *Global Tourism: The Next Decade*, edited by W. Theobald, 55–67. Oxford: Butterworth-Heinemann Limited.

Davis, F. 1973. *Yearning for Yesterday*. New York: The Free Press.

Donnan, S. 4 March 2015. World Bank Chief Warns of Surge in People Facing Resettlement. *Financial Times*. www.ft.com/cms/s/0/d3e5f2d8-c286-11e4-ad89-00144feab7de.html#axzz3Td1Tzrrn. Last accessed 6 March 2015.

Fried, M. 1963. Grieving for a Lost Home. In *The Urban Condition: People and Policy in the Metropolis*, edited by L. J. Duhl, 151–171. New York: Basic Books.

Gerety, R. 13 May 2013. Mozambique's Mining Boomtown. *Guernica*. www.guernicamag.com/features/mozambiques-mining-boomtown/. Last accessed 31 December 2015.

Guyot, S. and C. Seethal. 2007. Identity of Place, Places of Identities: Change of Place Names in Post-Apartheid South Africa. *South African Geographical Review* 89: 55–63.

Halbwachs, M. 1980. *The Collective Memory*. New York: Harper and Row Colophon Books.

Heidegger, M. 1996. *Being and Time*. (J. Stambaugh, Trans.). Albany, NY: State University of New York Press.

Human Rights Watch. 2013. *What Is a House without Food: Mozambique's Coal Mining Boom and Resettlements*. Washington, DC: HRW. www.hrw.org/report/2013/05/23/what-house-without-food/mozambiques-coal-mining-boom-and-resettlements. Last accessed 20 September 2015.

Kirshner, J. and M. Power. 2015. Mining and Extractive Urbanism: Postdevelopment in a Mozambican Boomtown. *Geoforum* 61: 67–78.

Lewicka, M. 2011. Place Attachment: How Far Have We Come in the Last 40 Years? *Journal of Environmental Psychology* 31: 207–230.

Light, D. 2004. Street Names in Bucharest, 1990–1997: Exploring the Modern Historical Geographies of Post-Socialist Change. *Journal of Historical Geography* 30: 154–172.

Lillywhite, S., D. Kemp, and K. Sturman. 2015. *Mining, Resettlement, and Lost Livelihoods: Listening to the Voices of Resettled Communities in Mualadzi, Mozambique*. Melbourne, AU: Oxfam.

Lubkemann, S. 2008. Involuntary Immobility: On a Theoretical Invisibility in Forced Migration Studies. *Journal of Refugee Studies* 21: 454–475.

McMichael, C. 2003. Memory and resettlement: Somali women in Melbourne and emotional wellbeing. Unpublished doctoral dissertation, University of Melbourne, Melbourne, Australia.

Miles, M. and A. Huberman. 1994. *Qualitative Data Analysis: An Expanded Sourcebook*, 2nd ed. Thousand Oaks, CA: Sage Publications, Inc.

Milgroom, J. and M. Spierenburg. 2008. Induced Volition: Resettlement from the Limpopo National Park, Mozambique. *Journal of Contemporary African Studies* 26: 435–448.

Mosca, J. and T. Selemane. 2011. *El Dorado Tete: Os mega projectos de mineração*. [El Dorado Tete: Mega mining projects]. Maputo: Centro de Integridade Pública.

Pedro, J. 2011. *Reassentamentos forçados: Dos impactos às oportunidades*. [Forced resettlement: From impacts to opportunities]. Master thesis, University Institute of Lisbon, Lisbon, Portugal.

Relph, E. 1976. *Place and Placelessness*. London: Pion Limited.

Republic of Mozambique. 2012. *Regulations for the Resettlement Process Resulting from Economic Activities, Decree 31/2012, 8 August*. Maputo: Council of Ministers.

Rose-Redwood, R., D. Alderman, and M. Azaryahu. 2010. Geographies of Toponymic Inscription: New Directions in Critical Place-Name Studies. *Progress in Human Geography* 34: 453–470.

Scannell, L. and R. Gifford. 2010. Defining Place Attachment: A Tripartite Organizing Framework. *Journal of Environmental Psychology* 30: 1–10.

Selemane, T. 2010. *Questões à volta da mineração em Moçambique: Relatório de monitoria das actividades mineiras em Moma, Moatize, Manica e Sussundenga*. [Questions on the Return of Mining Activity in Mozambique: Report on Mining Studies in Moma, Moatize, Manica and Sussundenga]. Report 6676/RLINLD/2010. Maputo, Mozambique: Centro de Integridade Pública (CIP).

Soja, E. 1989. *Postmodern Geographies: The Reassertion of Space in Critical Social Theory*. London: Verso.

Spitzer, L. 1999. Back through the Future: Nostalgic Memory and Critical Memory in a Refuge from Nazism. In *Acts of Memory: Cultural Recall in the Present*, edited by M. Bal, J. Crewe, and L. Spitzer, 87–104. Hanover, NH: University Press of New England.

Stal, M. 2011. Flooding and Relocation: The Zambezi River Valley in Mozambique. *International Migration* 49: e125–e145.

Strong, M. 2016. The influence of place attachment, aspirations, and rapidly changing environments on resettlement decisions. Unpublished doctoral dissertation, University of Maryland College Park, College Park, Maryland.

Trigg, D. 2012. *The Memory of Place: A Phenomenology of the Uncanny*. Athens, OH: Ohio University Press.

Tuan, Y. F. 1974. *Topophilia: A Study of Environmental Perception, Attitudes, and Values*. Englewood Cliffs, NJ: Prentice-Hall, Inc.

United Nations. 2014. *Human Development Statistical Tables*. New York: United Nations Development Programme. http://hdr.undp.org/en/data. Last accessed 24 October 2015.

Witter, R. and T. Satterfield. 2014. Invisible Losses and the Logics of Resettlement Compensation. *Conservation Biology* 28: 1394–1402.

World Bank. 2015. *World Databank, World Development Indicators*. Washington, DC: World Bank. http://databank.worldbank.org/data/reports.aspx?source=2&country=MOZ&series=&period. Last accessed 24 October 2015.

2 Hazardscapes

Perceptions of tornado risk and the role of place attachment in Central Oklahoma

Randy A. Peppler, Kimberly E. Klockow, and Richard D. Smith

It has long been known that individuals develop perceptions about risks based upon various cognitive, social, and cultural factors (e.g., Douglas and Wildavsky 1982; Slovic 1987; Fischhoff *et al.* 1993; Breakwell 2001, 2007; Wisner *et al.* 2004; Kasperson and Kasperson 2005; Kahan *et al.* 2012) that in turn influence decision making (e.g., Lord *et al.* 1979; Weinstein 1980; Gilbert and Wilson 2007; Gigerenzer and Gaissmaier 2011; Reynolds and Seeger 2012). In recent years, social and behavioral scientists have devoted increasing attention to the perception of, as well as response and vulnerability to, tornado risk (e.g., Comstock and Mallonee 2005; Schmidlin *et al.* 2009; Simmons and Sutter 2009; Sherman-Morris 2010; Schultz *et al.* 2010; Hoekstra *et al.* 2011; Dixon and Moore 2012; Nagele and Trainor 2012; Senkbeil *et al.* 2012; Sherman-Morris and Brown 2012; Simmons 2012; Suls *et al.* 2013; Ash *et al.* 2014; Ripberger *et al.* 2015; Trainor *et al.* 2015; Jauernic and Van Den Broeke 2016; Casteel 2016; Ashley and Strader 2016). An area of research that scholars have devoted less attention to is individual perceptions about places in which risks occur and the emotional ties people have toward those places. Klockow *et al.* (2014) describe this phenomenon, which emerged from interviews with survivors of the 27 April 2011 tornado outbreak in the southeastern United States. Their research found that while most people received tornado warnings and understood the gravity of the situation, they carried with them conflicting *a priori* notions about local tornado threats that influenced their perceptions of personal risk as the tornadoes approached. These notions mostly acted to give individuals false confidence that they would not be hit directly by a tornado. While some of the perceptions were found ubiquitous throughout the population (e.g., protective nature of hills), others were strongly correlated to particular places or nearby geographic features (e.g., cutting a new interstate highway through a previously hilly, wooded landscape). The research by Klockow *et al.* (2014) resulted in a categorization for reasons why people hold such false confidence including, place attachment, local (vernacular) knowledge construction, optimism bias, and the social amplification of risk.

In 1990, Larry Danielson examined tornado stories in the midwestern United States to better understand how tornadoes impact regional consciousness. He explains that severe weather fills a significant space in peoples' lives, whether it serves as a backdrop to daily activities or as an exceptional force with life-changing consequences (Danielson 1990). Moreover, he found that tornado stories are commonplace in contemporary oral tradition and play a significant role in communicating and addressing tornado risk. Danielson's research motivated Klockow *et al.* (2014) to examine the role of *place* in tornado risk perception. Their research began in the southeastern United States and then expanded to Central Oklahoma – a place notorious for frequent and violent tornadoes.

Previous research suggests (see Danielson 1990; Klockow *et al.* 2014) that sense of place (particularly the character of place) plays an important role in shaping risk perception and ideas about local tornado climatology. Is it possible that place-based optimism influences how prone people feel to risk? Do people in some places possess a false sense of security? What role do recent or well-remembered events play in shaping peoples' risk perceptions? Do some places attain a "more risk prone" status than other places, and if so, why? How will a greater understanding of answers to these questions influence severe storm readiness? In this chapter we examine, from multiple perspectives, the conceptions of risk people feel in their hometown as compared to nearby towns.

Central Oklahoma provides an excellent case study because, for as long as data records have been kept, there is no discernable pattern to tornado activity. With the exception of generalized tornado paths tracking from southwest to northeast, the area exhibits a random pattern of tornadoes. General consensus among meteorologists holds that tornadoes can hit anywhere in a given region with equal probability. Anecdotally, however, specific places in Central Oklahoma curry special attention with respect to tornado risk, and local residents have developed attitudes and perceptions that are at odds with existing scientific knowledge.

Place attachment and local knowledge construction

In 1992, Setha Low and Irwin Altman described place attachment as involving a complex interplay between emotions, beliefs, and actions centered on a place; living in a place becomes embedded in everything we do and is the foundation upon which we base our life's activities (Low and Altman 1992). More recently, Scannell and Gifford (2010, 1) define place attachment as "the bonding that occurs between individuals and their meaningful environments." Place attachment research often considers several dimensions including place identity (how a place affects an individual or group's identity) and place dependence (the stresses and strains that an environment imposes upon people). Place dependence is particularly relevant to this chapter because as Masuda and Garvin (2006) found, individuals who are farther away from a hazard may perceive greater risk from the hazard than those closer when the hazard threatens something they value about the place. For example, the construction of a new manufacturing center may disrupt a bucolic landscape that urban dwellers visit to escape the bustle of city life. Conversely, place dependence can lead people to perceive less risk for nearby hazards when they consider the contextual circumstances. For example, that same new manufacturing center may be less threatening when one considers the local employment opportunities it may create. The geospatial relationship between nearness to hazard and risk perception can therefore be very complex. Our investigation into the multifaceted intersection between place, place attachment, and risk perception is underpinned by numerous works especially Tim Cresswell (2015) and Maria Lewicka (2011).

Recently, a few studies have sought to understand the intersection between the characteristics of place and local weather phenomena or their aftermath. For example, Veale *et al.* (2014) describe a local wind in northern England that is produced by the interaction of the atmosphere with a particular physical landscape. This localized wind pattern was only discovered after scholars analyzed different narratives from a local culture group. Silver and Grek-Martin (2015) found that repeated cycles of disorientation and reorientation during recovery in the aftermath of a tornado provide residents with stronger feelings of connectedness and a deeper sense of community. And most relevant to this chapter are two recent studies (see: Smith and Cartlidge 2011; Hall and Endfield 2016) that articulate how specific locations and key landmarks can help place memories and serve as important "anchors" people use to make sense of their surroundings.

One of the most intuitive places that people are emotionally attached to is "home." As discussed in the introductory chapter, home is seen by most people as a safe haven that is free of risk and danger. A secure place. Peoples' concept of home can unfold at many different scales. Home can be the dwelling where one resides or home can be a larger place such as a hometown (the place where one feels comfortable and content, life is familiar). At the largest scale home can include one's native land (a.k.a. homeland).

Yi-Fu Tuan (1975) and Edward Relph (1976) have greatly enhanced our understanding of home. As they explain, home is filled with deep emotional feelings that can sway our decision, influence our actions, and shape our perspective on life (see also: Cuba and Hummon 1993). Because home is so important to us as individuals, our vision of home can become distorted from reality. When local residents share similar conceptions about a place, a local (or vernacular) knowledge of a place develops. Local knowledge refers to local, place-constructed ways of knowing *gathered through* observations, beliefs, emotions, culture, and actions that come from living in a place. Local knowledge can be grounded in fact, but more often it is based on anecdotal information that is shared throughout a population.

By whatever process it develops, local knowledge is understood to be a social product that helps people make sense of their surroundings, something that is culturally and geographically situated (Antweiler 1998). Wagner (2007) described local knowledge as a vernacular science knowledge (a metaphoric and iconic representation of scientific facts). We find this concept of local knowledge particularly relevant because it helps explain how people create their own notions or "climatologies" of tornado occurrence based on generalized observations as part of the lived experience. As Klockow *et al.* (2014) explain, the local knowledge people generated about their home place can shape how they perceive and react to tornado risk. If one has the belief that home is secure and free of vulnerabilities, it may lead to the development of unfounded feelings of safety and security.

Tornado town halls

Building upon the successful of previous research in Alabama and Mississippi, we set out to learn how local tornado knowledge in Central Oklahoma affects people's perception of risk. We knew that many long-time residents of the region promulgated the idea that their community was immune from tornadoes because something in the local environment protected them from danger. Based on existing data, the scientific weather community has sought to dispel these myths, but has had little success.

In fact, over the past 65 years there is anecdotal evidence to suggest that some "hot spots" of tornadic activity can be identified. The hot spots are places that have been hit by at least four tornadoes in the past 20 years. These recurring events dominate the memories of many people in the area. Our goal is to understand how such local knowledge influences perceptions of tornado risk. To do this we conducted three town hall meetings (large focus groups) with local residents in the three Oklahoma towns of Norman, Moore, and Newcastle during the Fall of 2012 (Figure 2.1). Each meeting lasted about 90 minutes and 35 people showed up to discuss tornadoes. We augmented the large focus group data with 63 surveys taken at the Norman National Weather Center Weather Fest in November 2012. A wide spectrum of local residents attended the town hall meetings and Weather Fest. The average age of participants was 46 years, with 33 males and 63 females. Fifty-five participants had college degrees, 16 had taken a college course, 15 ended their formal education at high school, and 12 did not identify their educational attainment. As a way to reach out to the community and enhance the public service component of the tornado town hall meetings, we invited select meteorologists, emergency managers, and news broadcasters to attend the three events. We found that the weather and risk management experts were genuinely interested in hearing people's stories because it helped them better understand local knowledge and thus develop methods to more effectively communicate with the at-risk population.

After gaining IRB approval, we developed a survey that measured the perceived level of tornado risk for each respondent's hometown. The surveys also included open-ended questions that asked individuals to explain their responses and draw on maps where they thought tornadoes were more likely to strike in their area. Furthermore, we asked residents questions about particular characteristics that might inhibit or encourage tornadic activity. For example, was there something geographical, structural, seasonal, or even spiritual element that might have an impact on where a tornado would strike. We found that having individuals write down their general perceptions on paper before the town halls' discussions helped draw out variability that might not otherwise been present during the large group discussion.

Because we understood that residents of Norman held a lower level of tornado risk for their city, we asked those residents more specific questions about tornado activity in their community. Specifically, we wanted to understand if residents felt Norman was at greater or lesser risk of tornadoes than the cities of Moore and Newcastle. Overall our sample included 98 individuals.

Figure 2.1 Map of Central Oklahoma.
Cartography by Jeffrey S. Smith

It is important to point out that all three cities are located 20 miles south of Oklahoma City and in close proximity to one another. In fact, the city limits of Norman and Moore share a common boundary along a two-lane road while Newcastle is separated from both Moore and Norman by the Canadian River. Even though all three towns share much in common physically, each town has its own tornado history, and each houses demographically distinct populations.

Overall, Norman is a relatively affluent, university town of about 120,000 residents. The city is about 85 percent white and has a median family income of about $63,000. Norman has had three tornadoes touch down within its city limits within ten years of the sampling period (2002–12). The first tornado formed on June 12, 2009 and caused light EF1 damage to a neighborhood on the east side of town. Another tornado touched down on May 10, 2010 on the southeastern edge of town. As it picked up momentum it exited the city toward the east and attained EF4 status on the far eastern fringes of the city limits. The third tornado passed directly through town on April 13, 2012, just prior to our study period, but was only rated as an EF1. Notably, several large tornadoes passed close to Norman during this period without hitting directly, including two tornadoes on May 24, 2011: an EF5 tornado that threatened the west side of the city but ultimately lifted near the river that borders Norman to its west, and another EF5 that threatened the south-central portion of the city but also lifted around a mile from the border.

Moore, located due north of Norman, serves as a bedroom community of Oklahoma City. Eighty-five percent of the town's 60,000 people are non-Hispanic, white and the community has a median family income of about $48,000. On May 3, 1999 the infamous EF5 tornado passed through the central business and residential districts of Moore leaving a long mile-wide path of destruction in its wake. Four short years later another long-track EF4 tornado passed through the business and residential heart of Moore. In the intervening years, several weaker tornadoes struck Moore, including two EF0s that hit the rural eastern fringe of town on May 9, 2004, and May 13, 2009; a longer-tracked EF3 that passed through the eastern rural and high-end residential portions of Moore on May 10, 2010; and a brief EF0 tornado affecting the rural western edge of town on May 24, 2011. Our sampling period concluded in the fall of 2012, but shortly thereafter, another EF5 tornado touched down within Moore city limits on the afternoon of May 20, 2013.

Newcastle is a smaller town located due west of Norman. The community is very spread out geographically. It is home to about 8,000 people, but is quickly becoming an upscale suburb of Oklahoma City. The town is about 85 percent non-Hispanic, white and has a median family income of $77,000. The town has a dedicated storm shelter adjacent to the elementary school that can provide protection for about 1,500 people. Newcastle has had some recent tornado near-misses, but until the long-track EF5 tornado of May 20, 2013 (just after our data collection period), none have touched down within the city limits.

From our experience, the use of town hall meetings was a remarkably effective method of understanding people's attachment to place. Both individually and as a large group, members of all three communities not only felt comfortable, but were excited, talking about their hometown. It was truly an effective way to get at people's emotional connection to place.

Perceived tornado risk among local residents

From the results of the three town hall meetings we found that people's perception of place-based risk is highly dependent upon location. Within the three communities, people in Norman feel relatively safer from tornadoes than residents in the other two cities (Table 2.1). Norman residents assigned their town a risk factor of 5.81 on a scale of 1 to 10 compared to 7.63 for Moore, and 7.08 for Newcastle (this difference

Table 2.1 Risk perception among town hall participants, 2012. Scale participants used to gauge perception ranges from 1 (lowest perceived risk) to 10 (highest perceived risk)

	Norman	Moore	Newcastle	Central Oklahoma
Norman residents (n = 43) risk perception of:	5.81	7.63	7.08	7.19
Moore residents (n = 23) risk perception of:	6.19	7.41	7.30	7.31
Newcastle residents (n = 7) risk perception of:	7.33	8.17	8.5	8.67

	West Norman	Central Norman	East Norman
Norman residents risk perception of:	5.49	5.38	6.50

Authors' tabulations

was significant at the 95 percent level). For Central Oklahoma as a whole, Norman residents rated the risk factor as 7.19. People in Norman saw Moore as the place with the greatest tornado risk, but their perceived difference between Moore and Central Oklahoma in general was not statistically significant. This suggests that the people in Norman see their town as a safe haven within a more dangerous region.

There was risk variability within Norman, as well. People who live in the eastern part of Norman feel less safe (6.50) than individuals in other parts of the city (west = 5.49 and central = 5.38). Recent events have likely influenced these perceptions. Indeed, some of the older folks in attendance told us stories about what happened decades ago as if the event was fresh in their minds. And while several strong tornadoes have touched down on the eastern side of Norman in recent years, that part of the city is physically separated from the commercial center and is more geographically remote. It is worth noting that eastern Norman is a lesser affluent part of the city and many residents have a less favorable opinion of that part of the city. The fact that residents from other parts of Norman regarded eastern Norman as more dangerous is interesting in light of these socio-economic divisions.

Although Norman residents rated their city as having a low risk level, residents in Moore and Newcastle gave Norman a marginally higher risk rating. Moore residents rated Norman's risk level at 6.19, while Newcastle residents rated Norman's risk of tornadoes at 7.33. Despite these higher absolute risk levels, Norman is still perceived by residents and non-residents alike to be the relatively safest place among the three communities.

Norman residents provided some interesting explanations as to why their city is perceived to be less prone to tornadoes. Among the most cited reasons was the protective force of the Canadian River that separates Norman from Newcastle. Tornado near-misses in May of 2011 may contribute to this perception, as two separate large tornadoes on May 24 approached the western and southern sides of Norman. Both tornadoes fizzled out as they reached the Canadian River. Curiously, when reflecting upon the small tornado that traveled directly through Norman in

April 2012, some participants were forced to rethink their logic given the fact that the tornado formed just west of the river and crossed the river as a tornado. In other cases, residents suggested that geographic elevation plays a role in where tornadoes form. The eastern side of Norman is located on higher ground than other parts of city, and therefore is seen as more vulnerable to tornadoes. A few people cited a Native American spiritual legend that asserts tornadoes avoid burial grounds. One resident said that this particular myth was taught in school, while another person said it was common knowledge and applied to other nearby towns.

The discussion above focused on the city of Norman. It compares how the people of Norman perceive the tornado risk within their community and how non-residents view Norman. The tapestry of place-based risk perceptions grows more complex when we consider residents' perceived tornado risk in other communities. In Moore for example, we found that residents saw their community as having higher tornado risk than the two other cities (Table 2.1). While residents of Moore thought their town was more risk prone than other towns, they also rated the risk level in their city a 7.41, while residents of Norman and Newcastle assigned it a rating of 7.63 and 8.71, respectively. The ratings Moore residents provided for themselves were consistent with how they rated Central Oklahoma in general (7.31). In sum, while everyone agrees that Moore is perceived to have the highest tornado risk in the immediate area, residents of Moore consider their city to be only slightly more dangerous than other parts of Central Oklahoma. Additionally, the residents of Norman and Moore both see their own community as less prone to tornadoes than how outsiders perceive their city.

Data gathered from the town of Newcastle provided us with the most striking results. Overall residents of Newcastle rated the tornado risk level for their community at 8.5 as compared to 8.17 for Moore and 7.33 for Norman (Table 2.1). This was the highest self-declared risk level among the three cities, and Newcastle was the only city to offer a higher risk rating for themselves than other cities gave it. Based on the data we collected, people in Newcastle feel the most vulnerable to tornadic activity. When asked to explain the logic behind their perceptions, residents provided a number of interesting responses. First, Newcastle is located along Interstate-44 which enters the city from the southwest and continues northeast to Oklahoma City. Residents believe that since the interstate runs through the community in the same prevailing directions that storms tend to track, it gives tornadoes an easy path to follow through the community. Second, residents indicated that Newcastle is located southwest of Moore. Since storms are perceived to have a consistent history of heading to Moore, Newcastle is vulnerable due to its relative geographic location. Third, people in Newcastle remarked that their city was on the wrong side of Oklahoma City. It is commonly believed that places on the downwind side of a major city are protected by the urban heat island and the man-made structures that disrupt the force of a storm. Newcastle, by comparison, is "unprotected" because it is located in an open area on the southwest side of the metropolitan area and therefore more vulnerable to tornadic activity. When combined, these three factors led residents to believe that their city was doomed for disaster.

Central Oklahoma survey

Because our town hall sample size was relatively small, it was difficult to assign statistical significance to our findings. Therefore, we employed a more comprehensive survey in March 2016 that seeks to understand the perspectives of an additional 463 people living in the Central Oklahoma region (Figure 2.1). The survey includes general measures of place attachment consistent with previous research including Raymond *et al.* (2010); Williams and Vaske (2003); and Lewicka (2010). Notably, this survey was conducted three years after the EF5 fringe-of-Newcastle-to-Moore tornado in 2013.

Preliminary results from our survey of 463 people appear to support what we learned from the town hall meetings. We asked residents a wide variety of questions, but one was particularly revealing. We asked, "Do you believe there are any landmarks or features that *prevent* tornado activity?" People who live east and north of Oklahoma City felt less prone to tornadoes because of the cluster of high-rise structures located in the downtown area. Residents reported that "high rise buildings," "buildings downtown," "clusters of buildings because they grow heat," and "tall buildings that make them [tornadoes] go different directions" were all identified with the benefits of living downwind from Oklahoma City. Indeed, overall respondents living northeast of downtown Oklahoma City assigned their community the lowest tornado risk at 5.6 on average (Table 2.2). By comparison residents who lived in other areas surrounding Oklahoma City rated their community with a higher risk value (6.2 for northwest locations, 6.4 for southwest locations, and 6.7 for southeast locations). It was interesting to discover that residents in the towns of Jones and Harrah (both located northeast of Oklahoma City) rated the risk level of their towns at 4.0 and 4.4, respectively.

In addition to the urbanized area of Oklahoma City, the Canadian River is the other prominent feature in Central Oklahoma that influences peoples' perception of tornado risk. Both the town hall meetings and the survey revealed that residents living near the Canadian River felt less vulnerable to tornadoes. Finally, the survey

Table 2.2 Risk perception among Central Oklahoma survey participants in selected towns and quadrants in relation to Oklahoma City, 2016. Scale participants used to gauge perception ranges from 1 (lowest perceived risk) to 10 (highest perceived risk)

	How vulnerable do you feel your town is to tornadoes?		
	Quadrant Average	El Reno	Yukon	Edmond
Northwest	6.2	6.4	6.1	6.1
	Quadrant Average	Choctaw	Harrah	Jones
Northeast	5.6	6.5	4.4	4.0
	Quadrant Average	Mustang	Tuttle	Newcastle
Southwest	6.4	5.4	7.7	8.6
	Quadrant Average	Moore	Norman	
Southeast	6.7	10.0	5.7	

Authors' tabulations

also indicated that a good number of residents in Norman felt safer from tornado activity because their community is located on higher ground or because of the historic Indian burial grounds.

We also asked residents of the Central Oklahoma region if there are "landmarks or features that attract tornado activity." Overwhelmingly, respondents cited highways and low-lying areas as more prone to tornadoes. In fact, Interstate 44 was the most cited feature that attracts tornadoes. It was commonly asserted that the interstate, which runs from the southwest to the northeast through the Central Oklahoma region, facilitates tornadoes and provides storms with an easy path through the area. Some respondents also mentioned Interstate 35 as a contributing factor to tornado activity. It was even mentioned that the relatively close proximity of Interstate 44 and Interstate 35 creates a "highway convergence." This area where these two highways run parallel is perceived to be particularly vulnerable to tornadoes. The fact that major roads and highways often are mentioned by meteorologists during tornado warnings may influence people's perceptions. Survey respondents in the southern half of the region also identified "valleys," "low lying and open areas," "the flat landscape," and "rivers and lakes" as noteworthy geographic features that seem to attract tornadoes.

Finally, we found that proximity to the city of Moore influences how vulnerable people feel toward tornadoes. People who live in the northern part of Central Oklahoma (a considerable distance from Moore) overall feel much safer than people who live in the southern portion of the region (closer to Moore). The major tornadic events that have unfolded in Moore over the past 20 years seems to have impressed upon people that Moore is a target for tornado activity and anyone who lives relatively close to Moore is at a greater risk of being hit by a tornado. This perceived pattern also influences people's tornado preparedness. According to survey responses, people who live in the northern part of the region are less likely to prepare for or heed tornado warnings than residents in the southern part (Table 2.3). Interestingly, residents who live in Moore are more likely to prepare for tornadoes than anyone else. Once bitten, twice shy.

Table 2.3 Likelihood of Central Oklahoma residents to prepare for tornadoes or heed tornado warnings, 2016. Scale participants used to gauge perception ranges from 1 (least likely) to 10 (most likely)

Likelihood to prepare for tornadoes	Northwest 7.0	Northeast 7.1
	Southwest 7.6	Southeast 7.7
Likelihood to heed tornado warnings	Northwest 7.6	Northeast 7.8
	Southwest 8.2	Southeast 7.8

Authors' tabulations

Conclusion

Overall, we have learned that perceived tornado risk varies greatly, even over relatively small geographic distances. Variance in peoples' risk perception stems in large part from place-based characteristics (e.g., topography and man-made features), but there is also mis*placed* optimism based on past events. Local knowledge influences people's tornado climatologies. Furthermore, we discerned that there is also a "media affect." When local television meteorologists use significant geographical landmarks such as rivers and highways to frame where the ongoing tornado risk is taking place, it influences how people perceive risk. Differences in residents' perceived tornado risk is a genuine concern for experts who are trying to serve and educate at-risk populations. By understanding how local knowledge is formed, experts can work to counteract misperceptions and misinformation.

This study also adds to our understanding of secure places. By focusing on risk perception we have turned the idea of secure places on its head. When it comes to natural disasters and risk assessment, we found that the community in which one lives is a surrogate for *home*. This stems from the notion that one's home is one's castle – a fortified space protected from the outside world where one feels safe and secure. Despite overwhelming scientific evidence to the contrary, we found that people rationalize their home as safe from tornadoes. It is people's deep-seated emotional connection to place that influences their perception of reality. As Klockow *et al.* (2014) assert, if one assumes home to be a place of refuge and stability, seeing it as vulnerable could be antithetical and emotionally destabilizing.

References

Antweiler, C. 1998. Local Knowledge and Local Knowing: An Anthropological Analysis of Contested "Cultural Products" in the Context of Development. *Anthropos* 93: 469–494.

Ash, K. D., R. L. Schumann, III, and G. C. Bowser. 2014. Tornado Warning Trade-Offs: Evaluating Choices for Visually Communicating Risk. *Weather, Climate, and Society* 6: 104–118.

Ashley, W. S. and S. M. Strader. 2016. Recipe for Disaster: How the Dynamic Ingredients of Risk and Exposure Are Changing the Tornado Disaster Landscape. *Bulletin of the American Meteorological Society* 97: 767–786.

Breakwell, G. M. 2001. Mental Models and Social Representations of Hazards: The Significance of Identity Processes. *Journal of Risk Research* 4: 341–351.

Breakwell, G. M. 2007. *The Psychology of Risk*. Cambridge: Cambridge University Press.

Casteel, M. A. 2016. Communicating Increased Risk: An Empirical Investigation of the National Weather Service's Impact-Based Warnings. *Weather, Climate, and Society* 8: 219–232.

Comstock, R. D. and S. Mallonee. 2005. Comparing Reactions to Two Severe Tornadoes in One Oklahoma Community. *Disasters* 29: 277–287.

Cresswell, T. 2015. *Place: An Introduction*, 2nd ed. Chichester: Wiley-Blackwell.

Cuba, L. and D. L. Hummon. 1993. A Place to Call Home: Identification with Dwelling, Community, and Region. *The Sociological Quarterly* 34: 111–131.

Danielson, L. 1990. Tornado Stories in the Breadbasket: Weather and Regional Identity. In *Sense of Place: American Regional Cultures*, edited by B. Allen and T. J. Schlereth, 28–39. Lexington: The University of Kentucky Press.

Dixon, R. W. and T. W. Moore. 2012. Tornado Vulnerability in Texas. *Weather, Climate, and Society* 4: 59–68.

Douglas, M. and A. Wildavsky. 1982. *Risk and Culture: An Essay on the Selection of Technological and Environmental Dangers*. Berkeley: University of California Press.

Fischhoff, B., A. Bostrom, and M. Jacobs Quadrel. 1993. Risk Perception and Communication. *Annual Review of Public Health* 14: 183–203.

Gigerenzer, G. and W. Gaissmaier. 2011. Heuristic Decision Making. *Annual Review of Psychology* 62: 451–482.

Gilbert, D. T. and T. D. Wilson. 2007. Prospection: Experiencing the Future. *Science* 317: 1351–1354.

Hall, A. and G. Endfield. 2016. "Snow Scenes": Exploring the Role of Memory and Place in Commemorating Extreme Winters. *Weather, Climate, and Society* 8: 5–19.

Hoekstra, S., K. Klockow, R. Riley, J. Brotzge, H. Brooks, and S. Erickson. 2011. A Preliminary Look at the Social Perspective of Warn-on-Forecast: Preferred Tornado Warning Lead Time and the General Public's Perceptions of Weather Risks. *Weather, Climate, and Society* 3: 128–140.

Jauernic, S. T. and M. S. Van Den Broeke. 2016. Perceptions of Tornadoes, Tornado Risk, and Tornado Safety Actions and Their Effects on Warning Response Among Nebraska Undergraduates. *Natural Hazards* 80: 329–350.

Kahan, D. M., E. Peters, M. Wittlin, P. Slovic, L. L. Ouelette, D. Braman, and G. Mandel. 2012. The Polarizing Impact of Science Literacy and Numeracy on Perceived Climate Change Risks. *Nature Climate Change* 2: 732–735.

Kasperson, J. X. and R. E. Kasperson. 2005. *The Social Contours of Risk, Volume I: Publics: Risk Communication and the Social Amplification of Risk*. London: Earthscan.

Klockow, K. E., R. A. Peppler, and R. A. McPherson. 2014. Tornado Folk Science in Alabama and Mississippi in the 27 April 2011 Tornado Outbreak. *GeoJournal* 79: 791–804.

Lewicka, M. 2010. What Makes Neighborhood Different from Home and City? Effects of Place Scale on Place Attachment. *Journal of Environmental Psychology* 30: 35–51.

Lewicka, M. 2011. Place Attachment: How Far Have We Come in the Last 40 Years? *Journal of Environmental Psychology* 31: 207–230.

Lord, C., L. Ross, and M. Lepper. 1979. Biased Assimilation and Attitude Polarization: The Effects of Prior Theories on Subsequently Considered Evidence. *Journal of Personality and Social Psychology* 37: 2098–2109.

Low, S. M. and I. Altman. 1992. Place Attachment: A Conceptual Inquiry. In *Place Attachment*, edited by I. Altman and S. M. Low, 1–12. New York: Plenum Press.

Masuda, J. R. and T. Garvin. 2006. Place, Culture, and the Social Amplification of Risk. *Risk Analysis* 26: 437–454.

Nagele, D. E. and J. E. Trainor. 2012. Geographic Specificity, Tornadoes, and Protective Action. *Weather, Climate, and Society* 4: 145–155.

Raymond, C. M., G. Brown, and D. Weber. 2010. The Measurement of Place Attachment: Personal, Community, and Environmental Connections. *Journal of Environmental Psychology* 30: 422–434.

Relph, E. 1976. *Place and Placelessness*. London: Pion Limited.

Reynolds, B. and M. Seeger. 2012. *Crisis and Emergency Risk Communication (2012 Edition)*. Centers for Disease Control and Prevention: U.S. Department of Health and Human Services, Atlanta, Georgia.

Ripberger, J. T., C. L. Silva, H. C. Jenkins-Smith, and M. James. 2015. The Influence of Consequence-Based Messages on Public Response to Tornado Warnings. *Bulletin of the American Meteorological Society* 96: 577–590.

Scannell, L. and R. Gifford. 2010. Defining Place Attachment: A Tripartite Organizing Framework. *Journal of Environmental Psychology* 30: 1–10.

Schmidlin, T. W., B. O. Hammer, Y. Ono, and P. S. King. 2009. Tornado Shelter-Seeking Behavior and Tornado Shelter Options Among Mobile Home Residents in the United States. *Natural Hazards* 48: 191–201.

Schultz, D. M., E. C. Gruntfest, M. H. Hayden, C. C. Benight, S. Drobot, and L. R. Barnes. 2010. Decision Making by Austin, Texas, Residents in Hypothetical Tornado Scenarios. *Weather, Climate, and Society* 2: 249–254.

Senkbeil, J. C., M. S. Rockman, and J. B. Mason. 2012. Shelter Seeking Plans of Tuscaloosa Residents for a Future Tornado Event. *Weather, Climate, and Society* 4: 159–171.

Sherman-Morris, K. 2010. Tornado Warning Dissemination and Response at a University Campus. *Natural Hazards* 52: 623–638.

Sherman-Morris, K. and M. E. Brown. 2012. Experiences of Smithville, Mississippi Residents with the 27 April 2011 Tornado. *National Weather Digest* 36: 93–101.

Silver, A. and J. Grek-Martin. 2015. "Now We Understand What Community Really Means": Reconceptualizing the Role of Sense of Place in the Disaster Recovery Process. *Journal of Environmental Psychology* 42: 32–41.

Simmons, K. M. 2012. The 2011 Tornadoes and the Future of Tornado Research. *Bulletin of the American Meteorological Society* 93: 959–961.

Simmons, K. M. and D. Sutter. 2009. False Alarms, Tornado Warnings, and Tornado Casualties. *Weather, Climate, and Society* 1 (1): 38–53.

Slovic, P. 1987. Perception of Risk. *Science* 236 (4799): 280–285.

Smith, J. S. and M. R. Cartlidge. 2011. Place Attachment among Retirees in Greensburg, Kansas. *Geographical Review* 101: 536–555.

Suls, J., J. P. Rose, P. D. Windschitl, and A. R. Smith. 2013. Optimism Following a Tornado Disaster. *Personality and Social Psychology Bulletin* 39: 691–702.

Trainor, J. E., D. Nagele, B. Philips, and B. Scott. 2015. Tornadoes, Social Science, and the False Alarm Effect. *Weather, Climate, and Society* 7: 333–352.

Tuan, Y.-F. 1975. Place: An Experiential Perspective. *Geographical Review* 65: 151–165.

Veale, L., G. Endfield, and S. Naylor. 2014. Knowing Weather in Place: The Helm Wind of Cross Fell. *Journal of Historical Geography* 45: 25–37.

Wagner, W. 2007. Vernacular Science Knowledge: Its Role in Everyday Life Communication. *Public Understanding of Science* 16: 7–22.

Weinstein, N. D. 1980. Unrealistic Optimism about Future Life Events. *Journal of Personality and Social Psychology* 39: 806–820.

Williams, D. R. and J. J. Vaske. 2003. The Measurement of Place Attachment: Validity and Generalizability of a Psychometric Approach. *Forest Science* 49: 830–840.

Wisner, B., P. Blaikie, T. Cannon, and I. Davis. 2004. *At Risk: Natural Hazards, People's Vulnerability, and Disasters*. London: Routledge.

Part II
Socializing places

3   Constructing sense of place through place-labelization in rural France

Hélène B. Ducros

Although we struggle to define it, we recognize heritage when we encounter it. Heritage sites feel like special places. Yet, they may be taken for granted as places that just happen to be. In fact, they are often the result of careful staging. As social constructions, they rely upon the values guiding diverse place actors into deliberate ventures motivated by economic, environmental, political, and socio-cultural objectives. Heritage is indeed "designated", based on what specific societies select for self-representation. The line between *designated* and *designed* is fine, both terms sharing a Latin etymology in *designare* (designate) and *signum* (mark, sign). This chapter addresses the purposeful and intentional actions behind heritage places in rural France, to show that they often are not the result of happenstance. It explores what the French call *mise-en-patrimoine* of place, which captures the processes implicated in the designation and marking of place as heritage, as well as the preservation and valorization of heritage sites through branding (Bénos and Milian 2013). By examining how rural heritage is valorized in France, the chapter provides an understanding of how sense of place is constructed through heritage designation strategies that cultivate place attachment and place identity.

To appreciate how cultural landscapes are selected, shaped, maintained, and staged as "heritage" and the impact of such designation on place itself, this chapter relies on an examination of the *Association des Plus Beaux Villages de France* (Most Beautiful Villages of France – MBVF). The mission of this grass-root organization is to revive village life by valorizing rural heritage through place-branding. Its model of place-making has reached beyond France to 13 other nations across three continents. This diffusion has accelerated since the new millennium as the association affirms its influence over rural landscape preservation and valorization, both nationally and globally. Through its branding scheme, the Association seeks to shape the perception of heritage sites and the personal and societal connection between people and rural places of patrimonial significance. By highlighting the ways in which "beautiful" is defined as to trigger emotional reactions to place and place stewardship, this chapter demonstrates how sense of place gets deliberately constructed through systematic and methodic recognition, designation, and assessment of heritage value, revealing how place-branding at the same time enhances local *typicity* and re-ignites a collective attachment to the rural.

Heritage-making, place-branding, and place-making

My inquiry into the MBVF places the organization at the intersection of heritage-making, place-branding, and place-making. *Heritage-making, patrimonialization* (Nora 1997; Chastel 1997; Hervieu 2012), or *heritagization* (Poria 2010; Smith 2006; Harvey 2001) describe the turning of the past and its tangible or intangible traces into "heritage" or "patrimony". Scholars have theorized on this omnipresent contemporary tendency (Gravari-Barbas 2005; Drouain 2006; Jeudy 2008; Heinich 2009), often expressing it as an ailment of modern societies (Lowenthal 2015) that traps the present and the future into *passéist* or narcissistic considerations and cripples us into barren nostalgia that produces a reimagining of the past that may stand in disconnect with the social history of the places it precisely seeks to represent.

Place-labelization[1] is both an instrument of exceptionalization and a community-building tool because a label at once separates and brings together the members that form its network. Although the practice has gathered momentum (Fournier 2014), scholars have yet to fully explore its processes and address the questions triggered by the application to place and space of theoretical concepts formerly elaborated for goods and services (Pike 2015). Labelization is a way to produce a "local" separate and different from other "locals" in a context in which globalization has caused a fear of cultural harmonization and homogenization. Faced with economic and cultural struggles, territories increasingly use labelization to push forward place characteristics they claim are uniquely local. In rural areas, labelization rests on massive flows of cultural tourists searching for a lost rural past and seeking memorable and differentiated experiences from place to place. From international to local labels, localities join networks of places (under different "labels") they feel acquainted with, connected to, and similar to in terms of cultural value and development aspirations because:

> in a competitive context where territories are increasingly establishing brands and proactively engaging in branding activities, places that opt out risk becoming invisible, ignored and left behind. . . . Space and place brands now occupy a significant part in the identity and image management elements of territorial development and play an integral role in transformative projects.
>
> (Pike 2015, 176–177)

Despite the fact that place-branding practices are increasingly attracting the attention of place entrepreneurs, community leaders, and researchers in various fields (often in an urban context), the literature on place-branding remains scarce and scattered (Govers and Go 2009), especially in the English-speaking world. Social scientists face the challenge of engaging with methods, publications, and a terminology typically reserved for the study of business and marketing. However, place-labelization has received more significant attention in the French literature, particularly focused on heritage labels and non-urban areas. The fact that the reflection over the role of place-labelization in territorial development has matured there reflects the reality that the phenomenon itself has already been part of the socio-cultural landscape for decades, with several prominent territorial labels operating since the early 1980s. Practically, place-labelization relies on local cultural resources, whether tangible or intangible. When successful, labels produce place-based symbolic capital that is used in economic development (e.g., tourism). But labels also matter in local identity production. In European rural development, local culture has been defined as intellectual property and rural identities used as assets in social and economic development (Ray 1998). "Cultural economy" (id.) or the "economy of culture" (Greffe 2005) are often invoked in territorial develop-ment as an "attempt by rural areas to localize economic control – to (re)valorize place through its cultural identity" (Ray 1998, 3). Place-branding is one element of the localization toolkit that brings together economic, social, cultural, political, and environmental considerations and objectives.

Place-making through labelization is understood as a dynamic and interactive multi-layered process encompassing reflection, selection, and exchange, rather than simple consumption of place (Govers and Go 2009). Through place-labelization it is the identity *of* place as well as one's identity *in* place that gets constructed by an array of actors who imbue place with meaning (Govers and Go 2009). For Blain *et al.* (2005), applying branding theory to the understanding of place construction may be problematic, although they note that the field of marketing points to three important functions of place-labelization, which include: 1) the identification and differentiatiation of place by way of a recognizable name, symbol, or logo, 2) the assurance of a memorable experience in place, and 3) the reinforcement of plea-surable memories of experience in place. Furthermore, place-labelization requires "brand equity", which may be enhanced by name awareness, brand loyalty, and quality perception (Govers and Go 2009). Brand coherence and intelligible com-munication bridge identity, image, and experience. Place-labelization is expanding globally. Places compete on many levels while people seek to reappropriate their identity, reclaim their history and reinvest themselves into the local. "Intensified globalization has made people footloose, . . . requiring an answer to the question: 'Where is the sense of place?'" (Govers and Go 2009, 16).

In France the media has fueled the current social infatuation with the country-side and the collective desire for the rediscovery of "rural." Because landscape patrimonialization recognizes the value of place and promotes the local, it gives rise to new development modalities and opportunities. Beyond contributing to the construction of territories, *patrimoine* has turned into a resource of territo-rial development which can separate and distinguish one place from another in a process of localization (Landel and Senil 2009). Heritage "blooms" to become the "yeast" necessary to support local rural projects (Husson 2008), providing an "active lever" to propel rural territories into the future (Bonnerandi 2005) and con-stituting a "metonymy of territories" (Landel and Senil 2009). Furthermore, rural heritage preservation has enabled a move away from an overemphasis on historic monuments and other extraordinary artifacts – usually recognized under top-down state structures – to the vernacular, the mundane, and the intangible. In France, the move from the preservation of "patrician" heritage to everyday twentieth-century places (Lowenthal 2015) and the expansion of the patrimonial field has led to the strengthening of *petit patrimoine* as the embodiment of local identities and lived experience (Nora 1997; Gravari-Barbas 2005; Husson 2008) that are rooted in place. These specifically local signature traits of familiar lived spaces of the past (e.g., washing pits, wells, or benches) become worthy of preservation and instruments of shared values from which the countryside is re-imagined.

Over the last decade in particular, the *village* has been central in the public imagination of what characterizes rural France. Epitomized by the media eager to provide a receptive audience with glimpses of "hidden" or "lost" France, the village (once forgotten and neglected) has become the place to watch. On television, in magazines, or books, the village sells. The reasons are multiple and complex, well documented by generations of social historians of agrarian France. The renewed interest for those small rural *communes*[2] has given an impetus to place-branding schemes such as the MBVF. While the activities of the Association have largely been ignored by mass media for the first 30 years of its existence, in the last few years this has changed. The MBVF is now increasingly being solicited as have rural territories in general, as important multifunctional spaces (Husson 2008), whether as tourism destinations or for relocating urbanites fleeing "modernity" and seeking alternative lifestyles away from the city. Patrimonialization is indeed no longer led by preservation or social reproduction alone, but is entangled with reflexivity resulting from the crisis of modernity (Landel and Senil 2009). Place patrimonialization produces social spaces where people self-identify by putting in place the genealogy of those before them. Faced with mass anonymity, people seek out distinctive assets, which they find in places that are promoted as heritage.

Methods

This chapter draws upon three-tiered ethnographic fieldwork. First, information was collected through interviews of Association staff and participant observation in Association activities. This allowed me to highlight its methodology in fostering senses of places in rural France through its definition of the "beautiful" and the implementation of tools for the selection of villages in its exclusive membership. Second, I interviewed the mayors of 22 member villages with the intent of understanding: what motivates the desire for labelization, what is expected from this local development strategy, and what challenges membership brings. Finally, I completed semi-structured interviews with more than 100 key actors and residents in five member villages to better grasp inhabitants' relationship with patrimonialized places and how place attachment may be fostered through place-labelization. The five case studies were chosen as to provide regional diversity as well as different levels of touristic appeal. When working in very small communities, confidentiality is paramount. In many cases, a single identifiable trait would easily reveal a person's identity. Therefore, this chapter ensures the promise of anonymity I made, which allowed for many rich conversations about how individuals form emotional bonds with their lived environment.

Association of the Most Beautiful Villages of France

The Association of the Most Beautiful Villages of France formally began in 1982. It was intended as a response to the decline of rural *communes* and their lack of prospect for the future. It is a book by Reader's Digest Selection (1977) that prompted Charles Ceyrac, mayor of Collonges-la-Rouge (Corrèze), to solicit fellow rural mayors to think together about the future of rurality and about how to best transmit rural heritage to future generations. The book, entitled *Les Plus Beaux Villages de France*, invited readers to reacquaint themselves with the "emotions of French *terroirs* they already knew" and to "find joy in discovering new ones". Along with a close friend who was a field geographer, planner, and territorial development practitioner, Ceyrac conceptualized his model of historical preservation in much the shape it exists today. Initially, the Association was composed of those mayors whose villages were featured in the book and who were interested in exploring an opportunity for local development (economic, cultural, and social) around the architectural and historical assets of their respective village. The valorization of these assets into resources would become the basis for the blueprint of the Association's mission of rural heritage preservation. After more than three decades, the MBVF gathers 156 members spread out across 14 *régions* and 69 *départements*. These municipalities have voluntarily applied for membership, as the Association does not solicit members. The Association estimates that across France, out of over 36,000 rural *communes*, only 6,000 villages are compatible with the aesthetic canons of rural France it has defined. This "elitism" is not denied by the Association. Instead, it explicitly defends a policy of exceptionalism. After all, "we are the *most* beautiful villages of France, . . . the superlative is key." It is in its definition of the "beautiful" that the Association becomes an agent of place-making. "We want to create places of emotions" and "we want people to find something in our villages that they don't feel in others. . . . It is this promise of excellence that drives our efforts and gives sense to our mission" (interview with MBVF 2012). How is excellence constructed and delivered? And how does it affect the relationship between residents and their lived space?

"Magical site", "place of emotion", "unique experience", "place of self-discovery", "site of quality", "atmosphere", "*commune* of exception", "regional exemplar" are but a few of the phrases used by Association officers and mayors of labelized villages to express what the label represents and what they strive to promote through its criteria of place development and spatial use. To deliver this vision, the Association relies on a central document first implemented in 1991. It is called the *Charte de Qualité*. Constituting the backbone of the Association's action, the Charte contains the rigorous criteria that provide concrete (people sometimes say "objective") content to the adjective "beautiful". The different elements of the Charte, however, reach far beyond simple visual features of the built environment, going deeper into place development and place performance in the long term. The 27 criteria in the Charte de Qualité go well beyond the listing of aesthetically typical and architecturally remarkable amenities. The built environment is only part of place value. Furthermore, it is through the assessment process (*expertises*) that the Association ensures that the superlative "most" remains at the heart of its search for excellence. How are "most" and "beautiful" made tangible on the landscape?

The creation of the Charte de Qualité marked a turning point in the mission of the MBVF. Prior to 1991, the Association was essentially focused on analyzing possibilities, gathering information, strategic planning, and the definition and identification of resources. After 1991, it started concrete implementation, initiated a marketing plan for the "brand," and focused on fostering better coordination between internal and external resources. Since then, the emergence of strict assessment criteria to define "beauty" (i.e., "*qualité*") is at the center of all Association's operations. In effect, the term "quality" (of place) represents the institutional way to capture the concept of beauty featured in the title-name of the Association. In fact, the term "beauty" is usually not invoked in the Association's narrative. Instead, the terms "quality", "worth", "value", "merit", "appeal", "significance", "deserving', and "excellence" are preferred to justify membership choices. The Charte is paramount in decision-making in the *Commission Qualité*, the body in charge of reviewing membership applications twice a year. True to its grassroot principles, the Commission is composed of mayors of labelized villages and a few advisors. They use the Charte as a guide for their decisions. In the words of two commissioners, "it's the Charte that decides" and "the Charte is number one, we only apply it." Hence, while "places of emotions" may seem an abstract concept, producing them entails concrete work for many actors of place development.

The flow of applications never stops. Candidates are villages that want to preserve or reinvent a rural identity they have lost over decades of abandonment, depopulation, and neglect from public authorities and residents themselves. "We want to remain visible" is a phrase I heard many times during interviews with mayors of labelized *communes*. Since 1991, 485 decisions have been rendered by the Commission Qualité. These include new applications (of which 82 percent have been rejected) and mandatory periodic re-evaluations of existing members. Indeed, the label is not permanent since member villages are re-inspected every six years to ensure their attributes remain in line with the Association's vision of rural heritage places. During the evaluation process, heritage quality and valorization efforts clearly take precedence over all other aspects. Nevertheless, all weigh in to convince the Commission that the village presents the desired long-term vision supported by local political will for this development trajectory. The Association's overall mission is not meant to be a quick commercial fix for difficult economic times in rural areas, but rather a mechanism for highlighting what the past can bring to the future based on cultural assets. The goal is for these assets to get noticed, then enhanced, and finally cared for in the long term. The assessment conducted by the field expert (there is only one) is composed of three parts: the tangible aspects of the built structures or natural sites,[3] the urbanistic and architectural quality of the village, and the amenities put in place to enhance the experience and use of place. "Urbanistic quality" pertains to village surroundings, size and homogeneity of the built area, and variety in walking pathways. The architectural quality involves the harmony and homogeneity of built volumes and façades, roof materials and colors, and openings (doors and windows), as well as the presence of symbolic ornamentation (e.g., a medallion above a door frame, an ornate balcony, a sculpted bench). The expert remains attentive to what is visible, but not exclusively. Also assessed are the potential for developing place appeal and efforts taken by the municipality to support this line of development in the future (whether preserving assets or creating new ones). For example, the existence of a *document d'urbanisme* (a non-mandatory zoning document for *communes*), is regarded favorably because it is proof of long-term commitment to place preservation. Also key in the evaluation is the implementation of amenities making the cultural landscape accessible and intelligible to visitors (e.g., interpretation and directional signage). Moreover, adequate *végétalisation* is taken into account (Figure 3.1), as is the presence of amenities and services for tourists, such as parking areas, street lighting, or a tourist office. Likewise, the burial of utility lines is a must, for visible lines are considered unsightly (Figure 3.2). During the assessment visit, the expert grades the village on each category of features and creates a photographic survey. Both are submitted to the Commission Qualité for deliberation. These reports form the basis for membership decision. The Commission also takes into account brand coherence ("is this village compatible with the others? Does it make sense?") and reputation ("will this village help or hurt the MBVF label") as well as its own role

Figure 3.1 Successful *végétalisation* in Le Poët-Laval illustrates the preference for local species and "sober" open-ground planting rather than plastic pots.

Photo by author, 2009

Figure 3.2 Burying electric lines is the cost of beauty in Charroux. For small communes, burying electric lines represents costly public works, but it is an investment in the landscape that pays high dividends in visitor perception.

Photo by author, 2012

in guiding struggling communes ("can the label give them the push needed to help them accomplish their goal(s)?"). For cases where the Commission sees unrealized potential, it may accept candidates "with reservation." In such cases, the label is received as a recognition of work well-done and an encouragement to go further in place management.

In managing public space, in addition to the Charte de Qualité, individual mayors may have recourse to other instruments initiated and implemented locally (not by the Association). Two such documents are the *Charte des Terrasses* (pertaining to restaurant and café terraces) and *Charte des Commerces et des Façades* (pertaining to shops and façades). These documents stem from growing concerns village mayors have over competing uses of public space. Their main objective is to ensure that the village remains functional for its residents while restricting the expansion of café terraces or unsightly commercial architecture (e.g., signage) that may alter the sense of place and jeopardize the village's membership in the Association.

As noted by a MBVF officer,

> these *chartes* have long been in existence in the urban setting. However, their presence in rural villages results from a new process of spatial organization there and the realization that more development and in particular uncontrolled wild commercial development is not always better.

Noting that these thematic "chartes" are meant to complement the Charte de Qualité, a mayor confirms that,

> we are probably going to go in the direction of these Chartes. . . . The Charte de Qualité remains an excellent statement of general principles, but in reality we need precise guidelines to set out rules of use that protect our resources and place excellence.

Hence, the Charte de Qualité is well understood to be a document that brings together localities around shared place ideals, but it is not meant to be a to-do list that dictates place development. Thus, it lacks the concrete application strategy guide that many mayors seek in the administration of their *communes*. Another limit of the Charte is the lack of explicit ecological consideration, at a time in which sustainable development occupies a prominent position in territorial development policy (Ducros 2017).

Living in a beautiful village

What does it mean to live in a MBVF? How do residents react to the Charte de Qualité? Interviews with residents of labelized villages reveal that although many of them are aware of the guidelines, just as many are either unaware or only vaguely cognizant of them. This leads to misunderstandings, which can engender resentment and conflict, or at the very least ambivalence towards the purpose of the

Charte. Although some residents reject the restrictions they are subjected to when making decisions over the use of their property, others embrace the guidelines as a way to care for their place of dwelling and better connect with a revoked rural past that puts them in touch with their rural ancestry (whether that ancestry is grounded in the particular village or another). When speaking with residents, this ambivalence is clear. For example, a long-term resident expressed his frustration by saying, "I can't do what I want with my own house. Even if I understand why this is, sometimes it goes too far". On the other hand, another resident asserts "I am glad people can't do whatever they want with their houses because they should realize it's more than their house, in a way it belongs to all the inhabitants of the village", adding that each house is

> like a symbol of the history of the village, and we are all responsible for keeping it tangible and visible. If people destroy that or tamper with it in irreversible ways, we don't have a village anymore; it won't feel like our village anymore.

Behind the clash in values relating to place, what appears here is the expression of different attachments to place as place represents a biography of lived experiences, at the individual or collective scale.

As residents react differently to aesthetic guidelines within their village, they also contend with growing numbers of visitors in various ways. The triangular relationship between place, residents, and tourists highlights that villages are first and foremost sites of exchange. Some residents enjoy sharing their village with visitors. For example, a permanent resident explained that she leaves her garden gate ajar, inviting people to come into her courtyard and chat with her. Another, recently settled in the village, described how much he has learned about the village from conversing with knowledgeable tourists. Others describe facing winter months with mixed feelings. While they appreciate the time when fewer visitors roam the streets of the village, they also come to miss them during the low tourist season. A resident mentioned that "even though it is nice when the village is ours again at the end of the summer, still it is always a nice feeling to see the first tourists come around by Easter time. We know spring is here." Inhabitants are aware that the label contributes to the touristic success of the village. One resident explained that,

> we live here and we are proud that people find it beautiful enough to make a detour to visit. . . . Tourists give a sense to this place for the outside world. We already know it means something. To us. But tourists, they prove that it means something, period. And, no label, no tourists.

The touristic encounter is made possible through the label and it enables the actualization of place attachment, by making residents reflect about the value of place and their connection to it. Indeed, some residents reported that since the village has been awarded the MBVF label, they started asking themselves questions about the place and began noticing things they may have overlooked before. They also started taking advantage of new amenities intended to facilitate the discovery of place by tourists. One resident expressed that, "one of my favorite places here now is the panoramic overlook. I never went up there before. But now there is a nice path." Another echoes this sentiment by saying, "I never noticed before what the big deal was about this place. But now that it's been nicely done, I can see it better, it's more evident. . . . And I find it really beautiful indeed." For many residents, the village is simply home. However, the landscape of home takes on an additional dimension for them when they attempt to perceive it as outsiders do. The label enables this dual perspective. Moreover, residents of labelized villages often purposely visit other labelized villages, learning to see their own village through implicit comparison.

Conclusions

Place-labelization is paradoxical. Simultaneously, it promotes the exceptionality of place, yet it also creates an ideal place type. Because France is very diverse in geographic, climatic, geologic, hydrologic, and cultural aspects, as well as local agrarian histories, it offers a great array of local specificity and cultural landscapes that it is difficult to standardize. This chapter provides an overview of the ways in which some places in rural France have been assigned value based on their individual character while at the same time contributing to creating an ideal of the rural. Specifically, it focuses on an actor of place-making in places of patrimonial significance. The *Most Beautiful Villages of France* truly acts as a "brand", in accordance with Blain *et al.*'s (2005) theorization of branding. The MBVF Association has developed the necessary instruments and methodology to identify and differentiate place. Through the Charte de Qualité, the MBVF contributes to the construction of a sense of place that is uniquely local, yet representative of a common ideal of rurality. Its name and logo are widely recognized throughout French rural landscapes since labelized villages must display a sign at the entrance of the village showing their membership (Figure 3.3). The MBVF aims at enabling a "quality" experience of place that is consistent from place to place under the brand, yet an experience that remains unique and differentiated. Like a brand, it seeks to build its credibility through the promise of excellence while it reinforces memorable experiences through its organization as a network of places. Indeed, when visitors travel from one labelized village to another within the network, the pleasurable experience is cumulative over time. Finally, brand equity is constantly and thoughtfully being built through efforts at coherent actions, although the value of the brand is difficult to assess concretely. In fact a MBVF officer wonders "what is our brand worth? We need to know." It is important to delineate how the Association has become a brand, because it is through place-labelization that the MBVF Association seeks to (re)make place and create or affirm a *genius loci* (i.e., the enduring particularities that give a place its unique identity, character, and personality) (Lowenthal 1985; Relph 1976) in rural villages that wish to remain economically, culturally, and socially significant.

Figure 3.3 Logo posted at the entrance of Eguisheim.
Photo by author, 2014

As place-labelization is in full expansion today, with more and more labels sprouting in the landscape, this study provides an understanding of how an established label has fostered the valorization of places in rural France and as a result shaped the image of the village as an ideal of the rural. The label designates places of encounter and revitalizes peoples' attachment to place. Beyond the protection of built heritage, it tells a human story and (re)creates spaces of sociability. While anchored in the history of place and the tangible elements that embody it, it constitutes a catalyst to open to the future places that have felt largely ignored by rural policies in the twentieth century. Although individual villages are assessed for place quality, it is a common vision of the rural that the MBVF implements. This gives coherence to the network under the brand, which acts as a federating agent of the rural. Through place-labelization, the Association achieves a paradoxical standardization of place uniqueness. Since the meaning of place is shaped by the intentionality of experience (Relph 1976), labelization also produces place meaning by providing intentionality through the promise of a similarly enjoyable experience from village to village. The root values of the Charte de Qualité shape local assets based on the recognition and enhancement of place specificity as a valorized resource.

For Relph (1976) sense of place cannot be reduced to the sum of its parts since place endures even when visual perception, topography, social and economic activities as well as its past meanings and present condition change. Thus, place becomes a paradoxical object of study where "real" components are easily identified through experience of place while others are so subtle that they are difficult to measure formally. Through connecting with the past, heritage imbues place perception with affect. The Association stands as a concrete response to the paradox of place that Relph (1976) evoked. At the same time as it puts the accent on the uniqueness of place perceived through direct experience of place, it attempts to deconstruct the nebulous cluster of experiential "feelings" into items that are measurable and observable separately from each other. While we receive place en bloc, the criteria identified by the Association to assess place quality disarticulate perception into separate constituents. Hence, the Charte de Qualité constitutes an important analytical tool to understand place production. The Charte de Qualité produces and promotes canons of place aesthetic. It transforms rural localities into spaces emblematic of the rural where identity, economics, patrimonial preservation, culture, and environment are linked in a complex process of introspection, social reflection, and exchange, and where past, present, and future are interwoven. As it marks the perception of the countryside and social imagination of the rural, it becomes an actor of territorial development and a producer of place. It promotes place assets, transforming and valorizing them into resources that can be used for a multi-layered agenda, based on values which are not necessarily shared uniformly across villages or within.

The Association activates the dual process of (re)discovery of place and personal self-discovery where the self is intimately linked to place in its tangible and intangible aspects. By seeking to inform and convince, the Charte de Qualité reveals narratives of place, comforts place identity, triggers place awareness, and encourages place stewardship. Moreover, it allows for multi-level encounters: inhabitants with experts, with visitors, with each other, or with place itself. It awakens memory of place, and buried childhood recollection of parents or grandparents. As a result, residents appropriate even those spaces that are created with tourists in mind, because those places also hold personal meaning or symbolic value. When faced with the formal technical criteria established in the various *chartes*, residents most often respond with informal vocabulary such as "beautiful", "home," "countryside," "alive," "live again," "agreeable," "practical or impractical," and "ancestors." In particular, some permanent residents report becoming more interested in their place of dwelling because they want to understand why others find it interesting. They also report that this fosters conversations among village residents. Often, it is conflicting narratives of place that create space for social encounters. Moreover, newly arrived permanent residents often convey their desire to be part of the stewardship effort in the village. While it would be inaccurate to suggest that all residents react positively to the label, a great number of residents who I interviewed admitted to drawing pride from the village being awarded the distinction, whether they overall supported the label or not. In some villages, the label has become so intertwined with individual identity and place identity that several mayors reported that if the label were not renewed after the periodic re-inspection, they would not be re-elected. This is true less because of the potential economic loss than because their constituents could not bear such personal and collective humiliation on the national stage. "*De France*" (of France) is an important part of sense of self for village residents, because it is the recognition at the national scale that heightens their sense of responsibility and care towards place.

Edward Relph (1976) sought to explain how sense of place and attachment to place are factors in the making of place and the caring for place, as well as how the character of places intersects with people's individual identity. By showing how the Charte de Qualité becomes a manifesto for place stewardship, this chapter suggests that labelization of place as heritage can be constitutive of both heritage and identity. While the Association's methodology appears technical (embodied in the almost technocratic Charte de Qualité), its narrative remains lyrical. Indeed, it aims to create "places of emotions" and most recently even places "for dreaming" (Plus Beaux Villages de France 2016). The Charte reflects this ambiguity, mixing visual criteria that are easily seen and measurable (e.g., parking amenities) with subjective criteria that are more difficult to define, such as "harmony." Integrating both registers of criteria in the same place development plan gives the document the illusion of objectivity, appearing globally as formal, systematic, and shielded from potentially biased value judgments. In reality, the Charte proposes a collection of cultural landscape values to guide the construction of experiential places. It shapes patrimonial assets into cultural resources by transmitting and affirming these values through place assessment. Thus, place-labelization constitutes an ensemble of technocratic methods that succeed at objectifying the subjective. As it facilitates the release of the spirit of place by fostering exchanges in place and self-reflection, it contributes to the emergence of socializing places.

Acknowledgements

With thanks to the *Association des Plus Beaux Villages de France* for giving me access to its activities and the National Science Foundation for its support (Grant #1202703).

Notes

1 *Label* is used instead of *brand* by actors of territorial development, thus I use *labelization* rather than *branding*.
2 While widely used, the term "village" has no legal existence in France. Instead, the appropriate term is "rural commune" (defined as a municipality with less than 2000 inhabitants). Many municipalities fall under this category, the last census revealing that out of 36,529 communes, 75% are classified as rural with 50% having less than 500 inhabitants.
3 It is required that there be at least two protected sites in the village (listed on the national inventory).

References

Bénos, R. and J. Milian. 2013. Conservation, valorisation, labellisation: la mise en patrimoine des hauts-lieux pyrénéens et les recompositions de l'action territoriale. *VertigO* special issue 16 DOI: 10.4000/vertigo.13631.

Blain, C., S. E. Levy, and B. Ritchie. 2005. Destination Branding: Insights and Practices from Destination Management Organizations. *Journal of Travel Research* 43: 328–338.

Bonnerandi, E. 2005. Le recours au patrimoine, modèle culturel pour le territoire. *Géocarrefour* 80: 91–100.

Chastel, A. 1997. *La notion de patrimoine, in Pierre Nora, Les Lieux de mémoire, Tome 1: 1433–1469.* Paris: Gallimard.

Drouain, M. 2006. *Patrimoine et Patrimonialisation: du Québec et d'Ailleurs.* Montréal: MultiMondes.

Ducros, H. 2017. Confronting Sustainable Development in Two Rural Heritage Valorization Models. *Journal of Sustainable Tourism* 25: 327–343. DOI 10.1080/09669582. 2016.1206552.

Fournier, M. (ed.). 2014. *Labellisation et mise en marque des territoires.* Clermont-Ferrand: Presses Universitaires Blaise Pascal.

Govers, R. and F. Go. 2009. *Place Branding: Glocal, Virtual and Physical Identities, Constructed, Imagined and Experienced.* New York: Palgrave Macmillan.

Gravari-Barbas, M. 2005. *Habiter le patrimoine: enjeux, approches, vécu.* Rennes: Presses Universitaires de Rennes.

Greffe, X. 2005. *Culture and Local Development.* Paris: OECD.

Harvey, D. 2001. Heritage Pasts and Heritage Presents: Temporality, Meaning and the Scope of Heritage Studies. *International Journal of Heritage Studies* 7: 319–338.

Heinich, N. 2009. *La Fabrique du patrimoine. "De la cathédrale à la petite cuillère".* Paris: MSH.

Hervieu, B. 2012. Préface. In *Nouveaux rapports à la nature dans les campagnes,* edited by F. Papy, N. Mathieu, and C. Ferault, 7–12. Paris: Quæ.

Husson, J.-P. 2008. *Envies de campagnes, les territoires ruraux français.* Paris: Ellipses.

Jeudy, H.-P. 2008. *La machine patrimoniale.* Paris: Circé.

Landel, P.-A. and N. Senil. 2009. Patrimoine et territoire, les nouvelles ressources du développement. *Développement durable et territoires* 12. DOI : 10.4000/developpementdurable. 7563

Lowenthal, D. 1985. *The Past Is a Foreign Country.* New York/Cambridge: Cambridge University Press.

Lowenthal, D. 2015. *The Past Is a Foreign Country Revisited.* New York/Cambridge: Cambridge University Press.

Nora, P. 1997. *Les lieux de mémoire.* Paris: Editions Gallimard.

Pike, A. 2015. *Origination: The Geographies of Brands and Branding.* Oxford: Wiley.

Plus Beaux Villages de France. 2016. *Point.com: La lettre des Plus Beaux Villages de France (42).*

Poria, Y. 2010. The Story behind the Picture: Preferences for the Visual Display at Heritage Sites. In *Culture, Heritage and Representation: Perspectives on Visuality and the Past,* edited by E. Waterton and S. Matson, 217–228. Farnham: Ashgate.

Ray, C. 1998. Culture, Intellectual Property and Territorial Rural Development. *Sociologia ruralis* 38: 3–20.

Relph, E. 1976. *Place and Placelessness.* London: Pion Limited.

Sélection du Reader's Digest. 1977. *Les Plus Beaux Villages de France.* Paris/Bruxelles.

Smith, L. 2006. *Uses of Heritage.* London: Routledge.

4 Exploring place attachment and a sense of community in the Chacarita of Asuncion, Paraguay

Jeffrey S. Smith

Increasing urbanization has long been perceived as a positive aspect of human development and the growth of cities is a story of increasing progress and prosperity (Florida 2014; UN 2014). Urbanization is associated with sustained economic growth, a reduction in poverty, greater employment opportunities, and more efficient distribution of social services. Each year an estimated 70 million people are added to urban areas throughout the globe (World Bank 2016). By 2030, more than 60 percent of the world's population is expected to be living in urban areas and most of the earth's population growth will occur in cities of the Global South (Parsons 2010).

Rural to urban migration accounts for the majority of urban growth. As people move to cities they congregate in areas where affordable housing is available. The newly arriving population typically seeks employment in urban manufacturing, but because they lack essential skills many former rural residents are relegated to chronically low-paying jobs (Conway 1985). Income constraints from reliable employment prevent most of the newly arriving residents from entering the formal housing market. In the Global South affordable housing is found in disamenity zones (a.k.a. slums and squatter settlements) where self-help housing dominates (Griffin and Ford 1980). In developing countries 70 percent of housing occurs progressively. Households acquire land through informal purchase or invasion and gradually improve the structure over 5 to 15 years (Ferguson and Navarrete 2003). "In most developing cities, decent and safe housing remains a dream for the majority of the population" (UN 2012, 3). Most self-help housing settlements are located on marginal or undesirable lands including steep slopes prone to severe erosion or along riverfronts where seasonal flooding is common.

According to reviews published by Dennis Conway (1985) and Susan Eckstein (1990), governmental policies towards squatter settlements have vacillated from alarm and antagonism to acceptance and accommodation. Over the past five decades there have been three main approaches to dealing with informal settlements. In the late 1950s and 1960s the growth of squatter settlements was met with strong resistance and forced removal. The call of the day was to bulldoze existing settlements and discourage further migration to the cities (Conway

1985). By the late 1960s and 1970s leading scholars including Charles Abrams (1966), William Mangin (1967), and John Turner (1967) rejected the prevalent concerns about squatter settlements and instead advocated for the expansion of self-help housing (Bredenoord and van Lindert 2010). They helped governments see the economic and social benefits of informal settlements (Portes 1971). In the 1990s the ground under the squatter settlements shifted again. The most common governmental approach was to condemn the lands and accuse the residents of unauthorized land invasion. Civic officials and investors once again regarded squatters, and the settlements they lived in, as obstacles to continued prosperity and economic development (Parsons 2010). With this mentality in mind, city leaders have pushed for urban development projects that would evict the violators, raze the informal settlements, and convert the land into a prime real estate investment with modern infrastructure and public recreation. Such projects are popular within the business community because they have the potential of enhancing a city's competitiveness, attracting outside investment, stimulating economic growth, and improving the overall quality of life for local residents (Desai 2012). In cities across the globe riverfront properties have been particularly attractive to urban development.

The governmental approach that started in the 1990s reinforces the long-held perception that squatter settlements are the source of many of the city's chronic problems. Because squatter settlements are notorious for housing the poorest of the urban poor, these informal settlements are maligned as dangerous places where filth, social deprivation, illicit activities, and criminal impunity prevail (Berner 2001; Dietz 1980; Eckstein 1990; Mangin 1967; Parsons 2010; Ulack 1978). Moreover, they are a source of economic drain since unemployment is high and educational levels are low (Mangin 1967). Squatter settlements are generally assumed to be the worst place in which a person can live (Owusu *et al.* 2008). By labeling the land as blighted and unsafe for human occupation, city officials can put in motion plans to relocate the residents to new lands and avoid the protests and social unrest such acts can create. Furthermore, because neighborhoods with high rates of poverty and crime are associated with a lack of sense of community (Fraser 2004), governments typically anticipate weak and disorganized resistance to relocation (Dietz 1980).

As squatter settlements have grown throughout the cities of the Global South, more scholarship is needed if we are to understand the characteristics of informal settlements and the people who reside in them. To date, most research has focused on the growing demand for affordable housing, quality of dwelling units, and the extent of support services. Few studies have focused specifically on the social qualities or on understanding the sense of community found within a squatter settlement. Using the residential space known as the Chacarita in Asuncion, Paraguay as a case study, this chapter illustrates how low-income residents have developed a strong, cohesive neighborhood with deep feelings of attachment to place. Using links to existing theory, I illustrate that, contrary to popular belief, the Chacarita possesses all of the hallmarks of a neighborhood with a strong sense of community.

The Chacarita settlement

The city of Asuncion was founded by the Spanish in 1537 along the banks of the Paraguay River. As Spanish presence and influence in the region increased, the colonial government devoted considerable attention to the Rio Paraguay since it was the main artery for exploration and trade. After Paraguay gained its independence in 1811, the government largely turned its back on the river because the annual flooding caused considerable damage and inconvenience for the local population. As a result, newly arriving immigrants to the city who lacked financial resources found the flood plain next to the city a convenient, affordable place to live. Not only was land free for the taking, but its proximity to the bustling downtown enabled residents the opportunity to secure employment (Borjesson and Bobeda 1964; Causarano and Chase 1991).

By the late 1880s the Chacarita was an established squatter zone on the edge of downtown Asuncion (Morínigo 2003). The term Chacarita refers to small farms where residents engage in subsistence gardening. The first settlers in the Chacarita were marginalized groups (e.g., Guarani Indians, blacks, and mulattos) who had difficulties finding reliable employment within the city. Over time two sections of the Chacarita formed. Within the flood plain is an area called *Zona Baja Inundable* (lower flood zone) and located on a small terrace slightly above the annual flood zone is *Zona Alta* (upper zone). As more and more families moved into the Chacarita the poorer and marginalized populations were shepherded toward the lower flood zone while people of higher social order (e.g., people of Spanish descent and longer-term residents) took to the upper zone (Morínigo 2003).

According to government documents and census reports, the Chacarita district is officially called *Barrio Ricardo Brugada*, yet the vernacular term has prevailed among the general public. In 1994 about 1,195 families lived in the Chacarita with an estimated population of 6,000 people. The 2002 census (the latest, published census available) indicates that Barrio Ricardo Brugada (Chacarita) had a population of 10,455 permanent residents (DGEEC 2005). When the 2012 census finally reaches print, census officials expect the Chacarita population to remain at approximately 10,000 people.

Downtown Asuncion and the Chacarita exhibit a symbiotic relationship. Because the Chacarita is located immediately adjacent to the downtown area, residents can walk to work and places for shopping, thus saving money on public transportation. At the same time, residents of the Chacarita fill important jobs within the government and downtown businesses including low-level office staff, security, public and private maintenance, janitorial and hospitality services, garbage collection, and landscaping. There is also a very large informal economy in parking security and car washing filled by Chacarita residents.

In her Master's thesis, Annalia Morínigo (2003) highlights that the Chacarita exhibits many of the characteristics typical of a squatter settlement. There is a general absence of formalized streets and nearly all of the homes are informally built using mixed, recycled materials (Figure 4.1). The area is underserved by government supported infrastructure. Electric power lines cross through the area

Figure 4.1 View of homes in the Chacarita with the Rio Paraguay and Argentina in the
background. The large structure in the center of the photo is an indoor fútbol
(soccer) stadium/dance hall.

Photo by author, 2014

allowing residents to tap into the grid, but most homes (especially in the lower
flood zone) lack reliable water, sewer, and sanitation. Furthermore, police and
fire protection services do not extend into the Chacarita. Residents in the area
have lower than average life expectancy and educational attainment, and high
unemployment as well as widespread chronic health problems. Despite access to
formal employment, many of the residents earn income informally by washing
windshields, selling handicrafts, street vending, and begging (Morínigo 2003).
Each year during the Rio Paraguay's annual flooding, residents of the lower flood
zone are forced to move their belongings to higher ground usually found in the
city's parks, parking lots, and alleys (Canese 2007). As the waters recede the families return to their homes.

In 2002, the government of Paraguay began looking at the land occupied
by the Chacarita residents as a potential site for urban renewal. Because living conditions are deemed deplorable and the area is commonly regarded as a
haven for criminals and illicit activities, the government sees the removal of
the Chacarita as in the best interest of both residents and the city (Gatta 2014).
In 2014 the government solicited bids from five architecture companies to help
them envision how the land could be developed. My time in Asuncion was

serendipitous for it coincided with when the five proposals were made available for public viewing. The intent was to solicit public comment on the project and the five proposals. When residents learned of the government's plans, protests erupted. News media covered the public demonstrations held by a small group of Chacarita residents. Currently, the situation is at a stalemate with neither side willing to concede. Residents of the Chacarita remain steadfast in their opposition to being removed and the city has moved ahead with improvements to the nearby *Costanera* (riverfront property) (e.g., levee, recreation facilities, and modern infrastructure). As of late 2016, the government has left the Chacarita and its residents undisturbed.

Methods

Data for this chapter originated from a wide variety of sources. To better understand the historical setting of the Chacarita settlement, I began with a thorough review of secondary literature including archival records, printed news reports, census data, journal articles, and student research projects.[1] With IRB approval, I began the ethnographic fieldwork during August and September of 2014. I derived the most fruitful sources of information from two guided field trips through the Chacarita[2] where I engaged in an analysis of the built environment and conducted semi-structured interviews with local residents of all ages including business owners, religious leaders, health care workers, and long-time residents. Additionally, with the help of two students from the Catholic University of Asuncion,[3] I conducted four focus group sessions. Each of these 30-minute discussions ranged in size from two to six individuals and explored the characteristics of the community and emotional connections that residents feel toward the Chacarita. Overall I gathered information from 47 individuals as I sought to understand the characteristics of the community and how connected people feel to the residential area. I documented my findings with an abundance of digital photographs, field sketches, and daily field notes. All data were retained in research journals and stored for future reference. The data collected enabled me to acquire a sense of how attached people feel to the Chacarita settlement and get a grasp of the sense of community and the characteristics of social connections people feel while living and working in the area.

Sense of community and place attachment

Social psychologists dating back to Emile Durkheim (*The Division of Labor in Society*) have sought to define and understand the concept of sense of community. Although a consensus definition across disciplines remains elusive (Fernback 1999; Freilich 1963; Hill 1996; Peterson *et al.* 2008), considerable progress has been made. In general, sense of community refers to a feeling of collective experience and common identity among a group of people. In the 1970s scholars began seeking to identify the elements that contribute to the formation of a sense of community. Numerous studies narrowed down the strongest predictors including length of residence in a neighborhood, level of satisfaction with the community or group, and personal connections to neighbors (Brodsky and Marx 2001; Glynn 1981; Riger *et al.* 1981; Long and Perkins 2003). The most cited work comes from David McMillan and David Chavis (1986) who identify four key elements contributing to a strong sense of community including *membership, social interaction, integration,* and *shared emotional connection.* From a review of the literature, it is apparent that there are strong correlations between sense of community and place attachment. The stronger the sense of community, the more attached people feel to a place (Lochner *et al.* 1999; Peterson *et al.* 2008). Drawing upon the sense of community literature (especially the authoritative work of McMillan and Chavis 1986), I identify examples within the Chacarita settlement that reflect a strong sense of community and speak to residents' deep emotional ties to place.

Membership

McMillan and Chavis (1986) explain that membership generally refers to how welcome individuals feel within the community or group. They identify five attributes, two of which are particularly relevant to this chapter. The first is established boundaries. Members of a community understand the social/cultural boundaries (e.g., proper dress, appropriate language, and rituals), because the accepted practices work to protect and reinforce the intimate social connections within the group. Less overt are the group's geographic boundaries. These boundaries are equally important because group activities unfold in designated places. When we enter the confines of a specific place, that place provides people with information and clues about who they are as members of the group. The boundaries of a place are a key ingredient that helps socialize individuals into the group (Rivlin 1982). For example, when people enter the confines of a church or mosque, they are immediately informed as to what behavior is deemed proper and acceptable within that space. Boundaries tell us who is welcome in the group.

The second attribute is a sense of belonging. Individuals possess a strong sense of belonging when their presence is reinforced and their thoughts and actions are rewarded. They feel like they belong and are legitimate members of the community. Evidence of these two attributes (boundaries and belonging) is readily apparent within the Chacarita settlement.

One of the things that stood out as I conducted my fieldwork was a clear understanding of the geographic boundaries that mark the edges of the Chacarita. On numerous occasions I attempted to enter the Chacarita as I walked down a street or alley. Each time I was turned away by residents who informed me that outsiders are unwelcome from entering the Chacarita without being accompanied by a permanent resident. Even police officers who stood guard at various formal entrances to the Chacarita told me that I should not enter the area alone because it is dangerous, but also because local residents forbid outsiders from entering without permission. Clearly, residents and non-residents alike know where the designated boundaries of the Chacarita are. The activity space is reserved for members only.

Residents of the Chacarita settlement also feel a strong sense of belonging. During both of my guided tours through the area, I talked with residents of all ages. Without exception, all of them expressed strong feelings of pride for their community; they knew they belonged. One anonymous woman approached me and said, *"Todos los que viven afuera piensan que este lugar es terrible. Amo este lugar. Está cerca del centro y conozco a todos mis vecinos. No me gustaría vivir en otro lugar."* (Everyone who lives on the outside thinks this place is terrible. I love it here. It's close to the downtown and I know all of my neighbors. I wouldn't want to live anywhere else.) Furthermore, numerous residents asserted that the negative reputation promulgated by the government and police stemmed from the bad behavior of a small group of the people dwelling there. As 65-year-old Alberto Urbieta told me:

> *Seguro que hay malas personas, pero la mayoría de los residentes son buenos, trabajadores y se ganan la vida honestamente. Son buenas personas.* (Sure there are some really bad people, but most of the residents are good, hard-working people who earn an honest living. They are good people).

They did their best to convince me that not only was the Chacarita a desirable place to live, but they were proud to be members of the community. While on the tour, I talked with Catalina Fernandez who told me,

> *Los ricos piensan que somos unos ladrones, pero no es así. Tenemos familia y amigos así como ellos, y trabajamos duro para mantener a nuestra familia, como lo hacen ellos. La única diferencia es que ellos tienen dinero y nosotros no. Eso no nos hace malas personas y este lugar aún así es buen lugar para vivir. Nos gusta vivir aquí.? Por qué deberíamos mudarnos a otro lugar?* (The rich people think we're a bunch of thieves, but that's not true. We have family and friends, just like they do and we work hard to take care of our children (pointing to her daughter playing nearby), just like they do. The only difference is that they've got money and we don't. That doesn't make us bad people and this is still a good place to live. We still like living here. Why should we move to some place else?)

More affluent households might find the housing and living conditions undesirable, but the people who live in the Chacarita possess a level of determination, resilience, and perseverance not always common among more affluent citizens. All of the residents I talked with exhibited intense feelings of pride in being members of the community. Each regarded living in the Chacarita as a privilege reserved for select people only.

Social interaction

According to McMillan and Chavis (1986), social interaction refers to the positive and rewarding feelings that members of a community have when they participate in group activities. As individuals get involved, interact with others, and cultivate personal relations they develop strong emotional connections to the community. There are countless venues where social interactions among a group of people can unfold, but some of the more common include social clubs, sporting events, fraternal organizations, and religious congregations. Within a group setting members of the community interact with one another and reinforce common ideas, values, and behaviors.

A common perception is that informal settlements lack social integration and group cohesion (Fraser 2004; Parsons 2010; Portes 1971). They are commonly identified with a transient population that develops few, if any, connections to others within the community. With this conception in mind, one of the most surprising elements I discovered within the Chacarita was all of the formal venues for social interaction that are available to local residents. While conducting my fieldwork I identified numerous small, elementary schools that offer educational services to nearby children. Each was informally built by local residents and locals told me that classes are taught by teachers who live within the Chacarita.

I also counted nine *capillas* (chapels) that offer a place to worship. Typically, these small buildings were established to serve as meeting houses for immediate residents only, but it was clear that considerable time, effort, and resources had been invested into establishing each of the chapels. The most impressive religious structure was the Santa Maria Goretti Catholic Church. Located within upper Chacarita, the building features a sanctuary that can hold at least 200 parishioners and is staffed by the local Catholic Diocese. In addition to regular mass services, each year the church offers numerous formal and informal social events that reach out to local residents and augment the sense of community within the Chacarita.

During my guided tours I also visited three dance halls and eight fútbol (soccer) fields/stadia. All three of the dance halls are impressive structures that host community social events and family celebrations (e.g., Quincineras and wedding receptions). Distributed throughout the Chacarita are ten fútbol clubs who compete against one another and help cultivate the skills of younger players who dream of turning professional. Each of the clubs is named after a prominent family who first settled in the area or a rural town from which a good number of the residents originated. Most of the eight fútbol fields are open air with limited seating, but two of them are covered, indoor arenas featuring two levels of seating with the ability to host several hundred spectators. The same stadia are multipurpose venues used for larger social events (e.g., concerts or community celebrations).

Unlike most of the housing units that are fabricated out of a mix of recycled materials, all of the schools, chapels, dance halls, and fútbol stadia are constructed of permanent building materials. Although they were informally built by local residents, each of them was durably built and some of the most attractive structures within the settlement. The Santa Maria Goretti Catholic Church is truly impressive. Moreover, the two fútbol stadia are so solidly built that they are used as temporary housing shelters during the annual flooding. It was evident to me that there is a strong sense of community through well-developed social interactions. All of these venues provide a place where local residents can interact and develop strong social connections. Despite the common misconceptions that outsiders have of squatter settlements, within the Chacarita there are numerous examples of social gathering places that reinforce a sense of community.

Integration

According to Kimberly Lochner *et al.* (1999) one of the cornerstones for a group of people to maintain a positive sense of togetherness is by working collectively toward a common goal. By combining individual efforts, the combined resources of the group help meet the needs of both the individual and the community. Not only does this social capital aid in community development, but it contributes to strong feelings of emotional connection. At all scales from small groups seeking to complete fund-raising campaigns to local governments engaging in public works projects, community improvement projects unite disparate individuals who work for the greater good. The integration process that unfolds during the completion of these projects reinforces membership within the group, builds a strong sense of community, and fosters attachment to place.

The common perception of informal settlements is that residents are so poor that all of their time and energy are focused on meeting their family's needs. As a result, there is little, if any, integration within the area. Within the Chacarita, however, there are a number of examples where members of the community have worked together to meet the needs of both the individual households as well as the community at large. One example is the small schools that pepper the settlement and provide educational services to many of the children within the area. Another example is the water and sewer systems that have been informally constructed within the settlement. In a piecemeal fashion residents of the upper Chacarita area have built a crude, but effective sewer system. Each family has a small pipe exiting their house that connects with a larger pipe that evacuates unwanted material to the river's lagoon. For the system to work properly, it is incumbent upon each household to keep their section of the main pipe free and clear of blockages (Figure 4.2). Likewise, in an attempt to address the annual flooding concerns, residents have built a crude gutter and water removal system that diverts water runoff away from individual homes so as to cause the least damage possible. It is remarkable that residents at the top of the hill slope have coordinated their efforts with households at the bottom. In both cases, members of the community have worked together to construct the systems and they must continue working together to keep them functioning as intended. Without a strong sense of community, such sewer and water diversion systems would be much less effective.

Shared emotional connection

McMillan and Chavis (1986) assert that the single most important element that speaks to a strong sense of community is a shared emotional connection. These are the emotional responses that members feel when they reflect upon their interpersonal connections. The most powerful way in which shared emotional connection is expressed is through shared history. When members of a community devote time and energy to recording the history of the group, it is clear that they possess a strong sense of community.

Figure 4.2 Resident of the Chacarita cleaning out the sewer connection between her house and the main line with a black hose.

Photo by author, 2014

Within the Chacarita settlement there are two excellent examples of shared emotional connection. Because the Chacarita has been used as a low-income residential space since the late 1800s, it is easy to understand how a shared history could develop. However, it must be remembered that the typical migration pattern involves families moving into the Chacarita temporarily as they seek secure employment and permanent housing within more attractive neighborhoods in Asuncion. In other words, once they can afford it, most residents exit the Chacarita in favor of nicer neighborhoods. The fact that there is a shared sense of history within the community is remarkable. One example of this shared sense of history can be seen at the community center. Outdoor murals and other artwork produced by local children and adults depict community leaders and key events within Chacarita's history (Figure 4.3). The large murals on the walls of the community center speak volumes to the sense of community found within the Chacarita. Likewise, the *Chacarita Museo del Barrio* (Chacarita Neighborhood Museum) is a second example of the shared emotional connection and shared history possessed by local residents. A visit to the museum reveals public documents, media articles, and various memorabilia dating back to the early 1900s. Each of the displays speaks to the shared history within the community. Many houses I entered also displayed information about significant events in the history of the Chacarita. For example, one house featured articles about José Flores, a former resident who grew up in the Chacarita and became a famous Paraguayan singer.

Figure 4.3 Murals on the west side of the Chacarita Community Center showing prominent community leaders, Santa Maria G. Catholic Church, typical community
activities, and the proximity of the Chacarita to the skyscrapers of downtown
Asuncion. The sign between the two murals reads Mural Mompox by Carmen
Mendoza Britos, 2011.

Photo by author, 2014

Summary and conclusions

Squatter settlements have long held the reputation for being havens of crime and
illicit activities, plagued by filth and abject poverty (Berner 2001; Dietz 1980;
Mangin 1967; Parsons 2010). It is also asserted that because residents of squatter
settlements are the poorest of the urban poor, they live in desperate conditions
where the majority of their time and energies are devoted to basic survival. As
a result, a common perception is that there is little sense of community among
residents (Owusu *et al.* 2008) and they possess no emotional attachment to place.
Their goal is to move out of the area as soon as they are financially able.

Beginning in the late 1960s a long list of scholars began questioning these commonly held perceptions and sought to shed new light on squatter settlements. With
the goal of better understanding the characteristics and dynamic qualities of squatter settlements, researchers have examined various aspects of informal settlements
including the process of land acquisition, quality of housing stock, characteristics of
shelters, provision of support infrastructure, and the economic links to other parts of
the city (Conway 1985; Eckstein 1990; Ferguson and Navarrete 2003; Florida 2014).

The resulting research finds that since the formal housing sector has failed to meet housing demands for all segments of society, households with limited financial resources are forced to engage in self-help housing (UN 2012). With more affordable land, rent, and housing than other parts of the city, informal settlements are an effective tool used by the urban poor to meet their housing needs (Eckstein 1990). Furthermore, residents fill important low-skilled, low-paying jobs that might otherwise go unfilled (Owusu *et al.* 2008).

Despite the evidence provided, many governments still do not recognize the important place that squatter settlements occupy within the urban fabric. Instead, officials commonly regard informal settlements and the people who live in them as drains on the larger economic system and obstacles to ongoing development and progress. It is a widely held assumption that if squatter settlements can be removed, then many of the social ills found within a city will be eliminated. One of the most common ways to justify removing a squatter settlement is to make room for an urban development project. After the squatters have been relocated and the land has been cleared, the business community is able to invest in real estate development. This helps the city stimulate investment, enhance its image, and improve overall quality of life (Desai 2012).

Because governments are under the false impression that squatter settlements lack a sense of community, they mistakenly believe that relocating the residents will be met with limited resistance. Building upon the sense of community literature (particularly the four-part model advanced by McMillan and Chavis (1986)), this chapter illustrates that residents of squatter settlements do indeed possess a strong sense of community and have deep emotional ties to place. Using the Chacarita settlement in Asuncion, Paraguay as a case study, and employing research methods grounded in cultural-historical geography (especially landscape analysis), this chapter showcases how the Chacarita exhibits elements of strong social cohesion.

Despite negative perceptions commonly held by outsiders, residents of the Chacarita settlement express considerable pride in living in the area. Their words and actions demonstrate that they feel like members of the community. One reason why residents possess such a strong sense of belonging is because of the high level of determination and perseverance required to overcome difficulties and survive in a marginalized community. Despite considerable neglect from the government, the families living in the Chacarita have established households and maintained a quality-of-life with which they are happy. Additionally, the fact that outsiders are forbidden from entering the Chacarita without an escort only adds to these proud feelings of membership within the community. The exclusivity of the Chacarita is similar to an elite country club or gated community where only authorized members are permitted. I found evidence that suggests residents actually help promote the dangerous reputation of the Chacarita so as to keep outsiders (especially government officials and land developers) from entering. The danger helps protect the solidarity within their community.

The strong feelings of membership are reinforced by the numerous places that foster social interaction. Members of the community have invested considerable time and resources into creating venues where individuals can come together and socialize. As Leanne Rivlin (1982) asserts, the interactions we have in a place help define who we are as members of a group. As revealed through my analysis of the built environment, some of the most important places where the process of socialization within the Chacarita unfolds include the schools, small religious chapels, dance halls, and fútbol stadia. Each place helps to reinforce the sense of community shared among the Chacarita residents.

Through mutual cooperation, residents of the Chacarita also demonstrate the ideas of social integration. By working together, residents have constructed crude infrastructural improvements (e.g., sewer system and water diversion) that not only help individuals, but aid the community as a whole. The positive experience residents feel when they work together and help maintain the system reinforces a positive sense of community.

As McMillan and Chavis (1986) assert, it takes time for a sense of community to develop. With the passage of time, members of a group cultivate relationships and form deep bonds with one another. Once a sense of community is established and members of the community develop trust through mutual support, then efforts are made to record the group's shared history. Within the Chacarita, the murals on the walls of the community center as well as the neighborhood museum provide two examples of efforts by local residents to record a shared history within the community. Both the artwork and museum allow members of the community to reflect upon their common origins and relish in their shared history.

This chapter speaks to one of the six types of places to which people become attached. The Chacarita is an excellent example of a socializing place because of the strong sense of community and feelings of belonging that permeate the settlement. Not only have residents developed strong relations with one another, but it is clear that residents of the Chacarita identify with and feel emotionally connected to the place.

Notes

1 Humberto Granada was instrumental in helping me piece together the historical background on the Chacarita.
2 Lillian Viera served as my guide. She and her family have lived in the Chacarita for more than 30 years.
3 Alejandra Rivarola and Mathias Baudelet were an amazing resource. Both played important roles in helping me gather information for this chapter.

References

Abrams, C. 1966. *Housing in the Modern World*. London: Faber and Faber.

Berner, E. 2001. Learning from Informal Markets: Innovative Approaches to Land and Housing Provision. *Development in Practice* 11: 292–307.

Borjesson, E. K. G. and C. M. Bobeda. 1964. New Concept in Water Service for Developing Countries. *Journal of the American Water Works Association* 56: 853–862.

Bredenoord, J. and P. Van Lindert. 2010. Pro-Poor Housing Policies: Rethinking the Potential of Assisted Self-Help Housing. *Habitat International* 34: 278–287.

Brodsky, A. E. and C. M. Marx. 2001. Layers of Identity: Multiple Psychological Sense of Community within a Community Setting. *Journal of Community Psychology* 29: 161–178.

Canese, M. 2007. Sociedad y Cultura Urbana: Convicencia Ciudadana en Plazas y Parques de Asunción. *Revista Internacional de Investigación en Ciencias Sociales* 3: 127–142.

Causarano, M. and B. Chase. 1991. *Mejoramiento Ambiental de los Barrios Chacarita y San Geronimo.* Asunción: CONAVI.

Conway, D. 1985. Changing Perspectives on Squatter Settlements, Intraurban Mobility, and Constraints on Housing Choice of the Third World Urban Poor. *Urban Geography* 6: 170–192.

Desai, R. 2012. Governing the Urban Poor: Riverfront Development, Slum Resettlement and the Politics of Inclusion in Ahmedabad. *Economic and Political Weekly* 47: 49–56.

DGEEC. 2005. *Atlas de Necesidades Básicas Insatisfechas.* Asuncion: Dirección General de Estadística, Encuestas y Censos.

Dietz, H. 1980. *Poverty and Problem-Solving under Military Rule: The Urban Poor in Lima, Peru.* Austin: University of Texas Press.

Eckstein, S. 1990. Urbanization Revisited: Inner-City Slum of Hope and Squatter Settlement of Despair. *World Development* 18: 165–181.

Ferguson, B. and J. Navarrete. 2003. New Approaches to Progressive Housing in Latin America: A Key to Habitat Programs and Policy. *Habitat International* 27: 309–323.

Fernback, J. 1999. There Is a There There: Notes toward a Definition of Cybercommunity. In *Doing Internet Research: Critical Issues and Methods for Examining the Net*, edited by S. Jones, 203–220. Thousand Oaks, CA: Sage Publications, Inc.

Florida, R. 2014. Why So Many Emerging Megacities Remain So Poor: How Globalization Has Changed the Nature of Urban Development. www.citylab.com/work/2014/01/why-so-many-mega-cities-remain-so-poor/8083/. Last accessed 11 October 2016.

Fraser, J. C. 2004. Beyond Gentrification: Mobilizing Communities and Claiming Space. *Urban Geography* 25: 437–457.

Freilich, M. 1963. Toward an Operational Definition of Community. *Rural Sociology* 28: 117–132.

Gatta, C. 2014. An Urban Project for Costanera of Asunción. *L'architettura Delle Città-The Journal of the Scientific Society Ludovico Quaroni* 3: 213–225.

Glynn, T. J. 1981. Psychological Sense of Community: Measurement and Application. *Human Relations* 34: 780–818.

Griffin, E. and L. Ford. 1980. A Model of Latin American City Structure. *Geographical Review* 70: 397–422.

Hill, J. L. 1996. Psychological Sense of Community: Suggestions for Future Research. *Journal of Community Psychology* 24: 431–438.

Lochner, K., I. Kawachi, and B. P. Kennedy. 1999. Social Capital: A Guide to Its Measurement. *Health and Place* 5: 259–270.

Long, D. A. and D. D. Perkins. 2003. Confirmatory Factor Analysis of the Sense of Community Index and Development of a Brief SCI. *Journal of Community Psychology* 31: 279–296.

Mangin, W. 1967. Latin American Squatter Settlements: A Problem and a Solution. *Latin American Research Review* 2: 65–98.

McMillan, D. W. and D. M. Chavis. 1986. Sense of Community: A Definition and Theory. *Journal of Community Psychology* 14: 6–23.

Morínigo, A. G. 2003. *Estudio Comparativo de Tipologias de Habitacion Para el Barrio Resistencia – Chacarita.* Master's thesis, Universidad Nacional de Asuncion, Asuncion.

Owusu, G., S. Agyei-Mensah, and R. Lund. 2008. Slums of Hope and Slums of Despair: Mobility and Livelihoods in Nima, Accra. *Norsk Geografisk Tidsskrift – Norwegian Journal of Geography* 62: 180–190.

Parsons, A. 2010. The Seven Myths of "Slums": Challenging Popular Prejudices about the World's Urban Poor. www.sharing.org/sites/default/files/images/PDFs/7_myths_report%20(2).pdf. Last accessed 11 October 2016.

Peterson, N. A., P. W. Speer, and D. W. McMillan. 2008. Validation of a Brief Sense of Community Scale: Confirmation of the Principal Theory of Sense of Community. *Journal of Community Psychology* 36: 61–73.

Portes, A. 1971. The Urban Slum in Chile: Types and Correlates. *Land Economics* 47: 235–248.

Riger, S., R. K. LeBailly, and M. T. Gordon. 1981. Community Ties and Urbanites' Fear of Crime: An Ecological Investigation. *American Journal of Community Psychology* 9: 653–665.

Rivlin, L. G. 1982. Group Membership and Place Meanings in an Urban Neighborhood. *Journal of Social Issues* 38: 75–93.

Turner, J. C. 1967. Barriers and Channels for Housing Development in Modernizing Countries. *Journal of the American Institute of Planners* 33: 167–181.

Ulack, R. 1978. The Role of Urban Squatter Settlements. *Annals of the Association of American Geographers* 68: 535–550.

UN (United Nations). 2012. Housing and Slum Upgrading. http://unhabitat.org/urban-themes/housing-slum-upgrading/. Last accessed 11 October 2016.

UN (United Nations). 2014. World Urbanization Prospects. https://esa.un.org/unpd/wup/Publications/Files/WUP2014-Highlights.pdf. Last accessed 4 June 2016.

World Bank. 2016. Urban Poverty: An Overview. http://web.worldbank.org/WBSITE/EXTERNAL/TOPICS/EXTURBANDEVELOPMENT/EXTURBANPOVERTY/0,,contentMDK:20227679~menuPK:7173704~pagePK:148956~piPK:216618~theSitePK:341325,00.html. Last accessed 29 May 2016.

Part III

Transformative places

5 Making place through the memorial landscape

Chris W. Post

In many ways the concepts of place and landscape may seem incompatible. Landscapes are largely analyzed as political materializations, forces of will that create a socio-economic norm based on capitalism (Cosgrove 1984; Harvey 1979; Johnson 1994; Leib 2002, 2004; Mitchell 2000, 2008). Places, on the other hand, are emotional and meaningful locales known through multiple bodily senses, commonly engendering positive feelings of attachment and validation (Agnew 1987; Casey 1996; Lewis 1979; Lopez 1990; Mitchell 2000; Ryden 1993; Sack 1997; Schein 1997; Tuan 1977). This chapter argues, however, that the realm of memorialization provides an opportunity to assess some overlap between place and landscape. I explore this potential using three particular commemorative landscapes.

As a scholar at Kent State University, my time on campus has provided me with many opportunities to consider the connections between place and landscape in a commemorative setting. Upon arriving at Kent I visited the May 4 Memorial, dedicated to the four student protestors killed by Ohio National Guardsmen in 1970. I was underwhelmed by the landscape, but could not identify why. Subsequent to my first visit, I have learned the details of both the shootings and the larger commemorative landscape that has evolved in dedication to the loss of life from that event. I have both published on this landscape dedicated to May 4 and served as a scholar consultant at the university's May 4 Visitors Center. Outside the Visitor Center and an annual vigil and memorial service, a truly critical commemoration of May 4 on the permanent landscape has not come to fruition. According to comments by the Kent State Board of Trustees, a former school president, and a survivor of the shootings, four markers in an active parking lot are the most affective memorials on the entire campus (Associated Press 1999; University Communications and Marketing 1999; Post 2016). Still, there is considerable controversy surrounding this piece of Kent State's permanent commemorative landscape. Important here is how critics complain that the location does not allow for a deep place-making experience.

Drawing inspiration from my experience and research at Kent State, I present here an analysis of three case studies that illustrate the possibilities for convergence between memorialization and place-making. To do so, I compare and contrast three memorial landscapes – the Vietnam Veterans Memorial, the Oklahoma City Memorial, and the Good Deeds Chairs memorial at Syracuse University.

Building upon methods first used by Perry Carter (2015), I examine the first two memorial landscapes utilizing review data from TripAdvisor and Google Reviews websites as they were available from March through May of 2016. I assessed how visitors comprehended their experience at these memorials. Specifically, I investigated comments pertaining to the roles of the visitors' bodies and those of the memorialized as components of these landscapes, in relation to the given memorial. This process gives voice to visitors without the discomfort of on-site interviews, perhaps making them more honest and/or critical of their experiences. It also allows visitors time to contemplate their experience and perhaps articulate their reactions more thoroughly. However, the collection of comments is a bit selective since reviewing in such a forum requires internet access, familiarity with TripAdvisor or Google Reviews, and an appreciation for sharing one's experiences to others. Despite these concerns, thousands of reviews are available for sites included on TripAdvisor alone, a website that allows for searching by keywords amongst the reviews, so that "scraping" the data is unnecessary. For the case study in Syracuse, New York, I coupled archival content analysis with landscape analysis conducted in October 2013.

Preceding these three case studies is a discussion of the role of the commemorative landscape, what it aims to do, and how it may use the human body as an element of its space. Following the case studies, I take a closer look at how memorialized landscapes enable visitors to empathize with the tragedy by connecting their body to the landscape which in turn may lead to greater feeling of place attachment.

Making place personal

The voluminous work by Ken Foote (2003, 2016) and Karen Till (2005, 2012) has greatly influenced my research. Both of these geographers have sought to understand the impact of tragedy on local communities, and how those communities respond through the act of memorialization. However, they take different approaches to understanding these processes. On the one hand, Foote (2003) emphasizes the landscape's importance in reflecting and reinforcing particular communal identities to tragedy, from "sanctification" to "obliteration." Till, by comparison, focuses more critically on the scale of the city, its politics, and the role of "memory work" (Till 2005) and how material memorialization, ceremony, art, and other community projects can create an "ethics of remembering" (Till and Kuusisto-Arponen 2015) and care for victims of tragedy (Till 2012). I assert that we can look more closely at the memorial space, how it represents the person(s) being remembered and its impact on the individual visitor. Thus, my focus is on the individual and the body – both of the visitor and the person(s) being remembered – as an active component of the memorial landscape.

Place production – that is, the transformation of nebulous spaces into meaningful places – requires affect between space and person and subsequent change in that person's ability to know and attach to the locale. In order for memorial landscapes to become places and sites of validation to visitors they must foster change within the visitor. Anne Buttimer (1976) notes the potential connection between affect and the opportunity to sense place. More recent scholarship on this issue proposes that developing a connection between the body and a commemorative landscape may best produce such attachment and the power, or agency, to empathize with the victims of violence, war, and tragedy (Drozdzewski *et al.* 2016; McFarlane 2011; Waterton and Dittmer 2014).

Locating the bodies of both the remembered and the remembering within the memorial landscape is, therefore, crucial to a deeper understanding of tragedy and the ability to cope with it. Paul Connerton (1989) discusses in detail how bodily performance produces memory. Though focusing mostly on the acquisition of knowledge and cultural reproduction, and not memorialization, Connerton's ideas can be extended into an analysis of the memorial landscape through his classification of bodily memory into the two camps of *incorporating practices* and *inscribing practices* (1989, 75). According to Connerton, inscribing practices are utilized to convey the meaning of a memorial – an explanation of what happened at a place. If, however, the body is *incorporated* into the landscape, then the reproduction of the socio-cultural lesson taught by that memorial landscape is more likely to occur.

Sociologist James Loewen (1999) provides a conceptualization of bodily relationship with his concept "hieratic scale." Here, Loewen connects the relative size (physical dimensions) and positionality (relative location) of both the person being commemorated and the visitor. Monuments such as the Lincoln Memorial, Jefferson Memorial, Mount Rushmore, among countless others, dwarf the body of the visitor, reinforcing the contributions of the remembered as they appear larger-than-life itself in comparison to the visitor.

Consider also the relative size and position between two figures in one memorial landscape. Loewen specifically makes the connection between non-whites compared to whites in commemorative landscapes using as example the Theodore Roosevelt monument outside the American Museum of Natural History in New York City. Roosevelt is shown on a horse above and in front of two indigenous persons. The size and position of the figures explains their relative and perceived social import. Visitors may not recognize or understand the elements of hieratic scale, but they are readily apparent to the trained eye.

Recently, Danielle Drozdzewski *et al.* (2016) edited a volume that focuses on the relationship between visitor and memorial space. The wide variety of essays therein underscore the importance of visitor agency and affect within the memorial landscape which bring the visitor into the space as much as they portray the victims of tragedy. Visitor and victim then meet in a middle ground of empathy, resulting in a virtual witnessing of struggle and loss embodied in the absence of the victim.

To further this approach of placing the body into the memorial landscape and the subsequent place-making experience, I contextualize three key concepts in my convergence of landscape and place in the commemorative landscape: *commemorative justice*, *public pedagogy*, and *witnessing*. Commemorative justice asserts that our commemorative landscapes should be meaningful, powerful, honest, and critical. As Derek Alderman and Joshua Inwood suggest, "Public remembrance of violence is a necessary tool for facilitating social compensation to victimized groups and moral education among the larger society" (Alderman and Inwood 2013, 194). Such sites should be inclusive and not merely reinforce some assumed normative narrative (e.g., "American Exceptionalism") or a benign mea culpa.

A significant component of commemorative justice is the production of a land-scape that serves to educate the general public. The landscape should not just merely normalize a nationalist story or that a tragedy happened, but it should focus on the social, political, and economic errors that led to that tragedy. A memo-rial need not be self-masochistic on behalf of its benefactors. Instead, memorial landscapes should be a form of public pedagogy where, through various media, the place reaches out to the public and informs them about their responsibility to uphold fairness and justice (Alderman *et al.* 2013; Giroux 2000; Loewen 1999; Post 2012; Sandlin 2010; Tyner *et al.* 2015).

Third, commemorative landscapes should encourage visitors to witness tragedy. In its most basic sense, anthropologist Pilar Riaño-Alcalastates states, "The work of witnessing establishes a communicative interaction between two individuals or collectives with one of the two addressing the other, and the other, in the act of transfer, becoming a listener and by extension, a witness" (Riaño-Alcalástates 2015, 285). Applying the notion of landscape as a discourse materialized (Schein 1997) empowers us to view the landscape as one of the parties in this affective interaction, and doing so then opens the visitor to a personal connection to the memorial landscape and its message. Riaño-Alcalástates further reinforces affect's significance by saying, "witnessing involves interaction, affective space" (2015, 285). Likewise, providing visitor agency (making the landscape more individual-istic and personal), increases the potential for visitors to internalize, witness, and learn from the tragedy being commemorated.

The larger question here is how any landscape may induce a visitor to bear wit-ness to a past tragedy. The first way is merely by its very existence. A known loca-tion can serve as a place of remembrance for people who personally experienced the tragedy being commemorated. Perhaps it even becomes a place of pilgrimage. As Till supports, "through the ritual of returning, one may experience a transforma-tive moment and confront personal social hauntings" (Till and Kuusisto-Arponen 2015, 15). Given their own internalization of the tragedy, a memorial brings back memories to a person already familiar with what happened, who can envision and even hear or smell the catastrophe themselves – their involvement in war, their loss of a family member, and other personal connections become paramount.

Witnessing may be experienced by more than just those who personally sur-vived tragedy. As feminist and media scholar Carrie Rentschler states, "In some important ways, then, the meaning of witnessing may shift from that of the particu-lar (first person) experience of the survivor toward the more generalized experi-ence of the media spectator" (Rentschler 2004, 297). Thus, a memorial landscape can position itself in a way that stimulates a sense of concern, empathy, and place validation. To this, the memorial landscape works as a kind of media that must be accessible to be meaningful. Location matters in assessing the role of a commemo-rative space (Alderman and Inwood 2013; Post 2015). Access via a fixed public location is one way to achieve this objective. However, a memorial landscape can also be mobile, bringing itself to the people and increasing the number of potential visitors, such as seen with the NAMES Project AIDS Memorial Quilt (Doss 2010).

A landscape may also be accessible to the body of the visitor, as I will illustrate below. In this sense, the body of the visitor becomes part of the landscape itself beyond the sense of vision, with opportunities to engage the senses of touch, hearing, and even smell. Such participation gives the visitor agency in the memorial landscape, personalizes the lessons learned, creates a sense of public pedagogy, and may lead a visitor to witness the loss of life and the importance of the event.

The concepts of commemorative justice, public pedagogy, and witnessing put the visitor at the center of the commemorative landscape by focusing on their agency and the ability of the landscape to make a difference with that visitor. Deeply laden with intense emotions, the goal of a memorial landscape is to change the visitor's perspective and enable them to connect to the tragedy. That experience (an emotional conversion) helps transform that memorial space into a place for the visitor and they may then become attached to it through this new emotional experience and commitment. Such experience validates the grieving individuals and groups through a sense of place. The following three examples, discuss how memorial landscapes may be produced in ways that enhance such transformations.

The Vietnam Veterans Memorial

The Vietnam Veterans Memorial on the National Mall in Washington, D.C., was designed by Maya Lin in 1981 and dedicated in November, 1982. Though initially viewed with derision by many, it has become one of the most respected commemorative landscapes in the United States. I focus here on The Wall of the Memorial, and not The Three Soldiers or Vietnam Women's memorials that evolved as compromises to construct the Wall (Wagner-Pacifici and Schwartz 1991; Savage 2009). The primary feature of the Wall is the listing of 58,307 names of soldiers killed and missing in action from the Vietnam conflict, defined as starting in 1957. It also includes the names of those killed by injuries sustained in Vietnam if the person died shortly after returning to the U.S.

Upon entering the space of the Wall, the visitor's body is given immediate agency. First, the polished gabbro reflects the visitor's image and those of several other monuments on the Mall as they become visible parts of the memorial (Figure 5.1). Second, the 58,307 names etched in the Wall give presence to the missing bodies of those killed and missing in action from the war. Guests frequently rub the names of their memorialized friends and family using pencils or crayons and paper, personalizing and connecting to not only their kin, but also the memorial at large. Third, as visitors walk along the Wall – the names are listed in chronological order of the enlistees' deaths – the path slowly dips down toward the apex of the structure (a 125-degree angle) and re-ascends toward the opposite end of The Wall. This motion symbolizes the moral descent of violence during war and the struggle to climb out and end it. Fourth, a number of objects have been donated to the memorial over time, reflecting what Miles Richardson calls "the gift of presence" (2001). Not merely gifts, these contributions to the landscape represent a piece of the visitor being left at the memorial. Flowers, personal possessions, and even a Harley Davidson motorcycle have been left at the Wall by visitors (Savage 2009). Combined with the rubbing and "taking" of names from the Wall, the visitor becomes part of a special dialectic of place that connects them not only to the memorial but also to the Vietnam War's impacts on life and American society.

Figure 5.1 Visitors read names on the Vietnam Veterans Memorial while the Washington
        Monument is reflected by the polished gabbro.

Photo by author, 2010

According to data I collected through reviews in Google and TripAdvisor, these
points combine to give the visitor agency beyond passive reflection – they become
part of the memorial and actively witness the tragedy and human cost of war. According to two reviewers, the Wall's reflection was of utmost note. "For those of us who
grew up in that era, the reflections of our faces on the polished black granite [sic] say
more than our tongues can express," said one visitor from D.C. "Furthermore in the
marbles [sic] reflection I was able to see the moon and Washington monument which
added a little bit of extra edge to the experience," said another visitor.

Engaging the memorial through walking was noted by nearly 13 percent of
the 6,387 reviews I perused on TripAdvisor. The walk down and back up was
called "moving" and "emotive." One reviewer added the memorial looked like "a
wound" from above the walk.

The act of witnessing was a common reference in reviews. Many writers
referred more to their witnessing of a veteran or other visitors leaving a gift or
finding and/or rubbing a name. A few, however, did note their own personal witnessing of the tragedy through their experience at the memorial. As one visitor
from Salisbury, Maryland, noted, "Intensity of the emotions when you witness
all those names . . . all of these do make it my favorite memorial in DC." Another
added, "Will definitely visit again, and bring my children to witness."

Personalizing the memorial through physical contact with the Wall was another theme in many reviews. As one reviewer remarked, "If you visit this hallowed place, touch these names, and remember the lives and honor of those Americans who did not return." These reviews validate what the memorial sets out to accomplish – to involve the visitor and engage multiple senses – hopefully with the result of bearing witness and making future peace.

The Oklahoma City Memorial

At 9:02 am on April 19, 1995, domestic terrorists Timothy McVeigh and Terry Nichols conspired to destroy the Alfred P. Murrah building in downtown Oklahoma City using a truck-contained bomb that killed 168 people, injured 680, and caused approximately $650 million in damage. The Oklahoma City Memorial was dedicated to the victims of this tragedy on April 19, 2000, and the adjoining museum opened one year later. The primary components of the outdoor memorial (the focus here) include bronze gates at the western and eastern ends of the former building (Figure 5.2), a reflecting pool between these gates, the "Survivor Wall" remnant of the Murrah Building, bronze chairs representing each of the victims killed by the bombing (Figure 5.3), and the "Survivor Tree" American Elm that, though damaged, still lives and thrives in its 100+ year old location.

Figure 5.2 The eastern gate of the Oklahoma City Memorial with the time of 9:01 (the moment before the bombing) reflects in the pool below.

Photo by author, 2014

Figure 5.3 Empty chairs that are lit at night to commemorate the victims of the Alfred P.
       Murrah Building bombing.

Photo by author, 2014

Most striking about the memorial landscape are the 168 chairs made of bronze
with glass bases that contain a light that shines at night. These individual markers are arranged by the floor that each victim was on at the time of the explosion.
Three unborn children are listed with their mothers. Nineteen smaller chairs mark
the second floor location of a daycare center in the building. These chairs, similar
to the names engraved in the Vietnam Memorial, make present the missing bodies
of those people killed in this tragedy. Expressing this absence by using a common
object (a chair) makes the loss of life palpable, personal, and unavoidable. It should
be noted that, though they may walk on the lawn where the chairs are located, visitors are prohibited from sitting on the chairs. Thus, the Oklahoma City Memorial
lacks the physical contact with the landscape that the Vietnam Veterans Memorial
allows. The chairs look out onto a reflecting pool that, similar to other pools and
even the Wall in D.C., simultaneously brings the visitor's body into the memorial
space. The pool is flanked by the gates displaying the times of 9:01 and 9:03, the
last moment of peace before, and the first moment of recovery after, the bombing.

TripAdvisor reviews highlight the emotional impact of the chairs, which were
mentioned in 558 of the 2,517 reviews I perused. As one reviewer from Charlotte,
North Carolina said, "The chairs are a striking representation of the lost lives."
Other reviewers noted visiting at dusk or night to see each chair lit. As one local

visitor recommended, "The best time to visit is at dusk or right after dark . . . it [sic] is beyond moving to see all the chairs, which represent each person who died, lighted." Another local visitor reinforced, "As you see the empty chairs, each one signifying an extinguished life, you can't help but be moved by the loss."

Bearing witness at the site, particularly in the museum was also a focus of several reviews. As one visitor from Colorado wrote, "Be prepared to travel back in time and witness the horror of terrorism. You start your journey as one would a normal day in Oklahoma City . . . but are thrust into a world of chaos and commotion." A visitor from Texas said similarly, "The museum gives witness to the triumph of good over evil and the strength and power of the human spirt!" The memorial in Oklahoma City is clearly an affective landscape allowing the visitor agency to learn and bear witness to the bombing and its impact on the nation.

Good Deeds Chairs and Syracuse University

Every year Syracuse University constructs a temporary memorial toward the Pan-Am Flight 103 bombing over Lockerbie, Scotland. On December 21, 1988, Libyan terrorists destroyed the plane on its leg from London to New York City. On the flight were 35 Syracuse students on their way home from a semester in London. In 1990 Syracuse built a memorial wall – the Place of Remembrance – at the front of its campus that lists the names of the victims. However, here I focus on the relatively recent, and temporary, Good Deeds Chairs memorial due to its connection to the theme of incorporating visitors into its landscape. Every fall Syracuse hosts *Remembrance Week*, a series of events that includes a vigil service and the announcements of 35 annual Remembrance Scholars. The week before these event organizers place 35 chairs around campus. Students are encouraged to write on the chairs listing good deeds they have done for, or will "pay forward" to, others. Over the weekend, these chairs are moved to the school's main quad (Figure 5.4). Outfitted with student acts of goodwill and blue doves, the chairs are arranged on the quad as the students were seated on the airplane – bringing the visitor to the tragedy and once again echoing the role of the body in the memorial. The visitor's body is present amidst directly absent victims' bodies.

Though Google or TripAdvisor reviews of this site are not available, due to its on-campus and temporary nature, comments in various local media outlets reinforce the power of this memorial. Karolina Lubecka, a 2014 Remembrance Scholar noted the chairs were one of the most powerful displays on campus since students will have to walk by and find out what they are about (Kilgannon 2014).

In 2012 Liz Mikula, also a Remembrance Scholar who designed the logo for that year's Remembrance Week and drew the logo on each chair, told the *Daily Orange* student newspaper,

> It unifies everyone on campus; it relates to everyone on campus. . . . It is important to be a part of something bigger than yourself on campus. Our message, look back and act forward, is an important one a lot of people can take to heart.

(Freundlich 2012)

Figure 5.4 Good Deeds Chairs on Syracuse's Quad in 2013 representing each life lost in the Lockerbie Bombing sit on Syracuse's Quad.

Photo by author, 2013

Alise Fisher (another Remembrance Scholar) added, "A lot of people know to a certain extent about Remembrance Week, but don't understand why it's relevant to them. This is a nice gateway to capture interest in a small way to get involved in a big way" (Freundlich 2012). Though not directly using the word "witness," these two scholars certainly point toward the immediate embodiment of remembrance that these chairs bring to campus for all students. Such embodiment allows the students to realize their opportunities to continue what was taken from the victims of Flight 103 – to earn a college degree and contribute to the betterment of society – to live their lives. What is more, the memorial space provides students an opportunity to engage with their communities and actively work for justice and peace through volunteerism and everyday work.

Making place through a corporeal experience

In all three of these memorial landscapes, the body plays a central role in helping visitors achieve a heightened understanding of the tragic events. The human body is key to this witnessing process. The bodies of the victims may be physically absent but they are accounted for in a material way through the names on the Wall or in the spaces occupied by chairs. Such affective landscapes, as described in the examples above, possess the opportunity to induce a real connection with visitors, if they are actively engaged and open to remembering. Each of these memorial spaces becomes a place where visitors have the opportunity to develop strong emotional connections rooted in empathy.

Thus, the body of the visitor becomes integral to the memorial's message that reminds one of the delicate balance between life and death. This dialectic of absence and presence provides visitors the chance to recognize their relationship with the commemorated victims. Though many memorials achieve this by listing the names of the remembered, I suggest that a mere reading of names (Connerton's "inscription" (1989)) does not actively reinforce this dialectic. As Judith Wasserman elucidates, "really good memorial places allow visitors [to gain] experiential insight . . . [by] viewing and touching artifacts, moving in ritual patterns, and engaging in community activity" (1998, 43). It is especially through multi-sensory stimulation (e.g., seeing one's reflection in a wall or pool, or interacting with an empty chair) that visitors are confronted with the loss of human life.

Analytically, this dialectic of presence and absence performs two things that may create a unique place attachment between visitor and memory place. First, the visitor becomes decentered. In his book *The Betweenness of Place*, J. Nicholas Entrikin writes:

> Associated with this transformation is our greater awareness of the funda-
> mental polarity of human consciousness between a relatively subjective and a
> relatively objective point of view. The former is a centered view in which we
> are a part of the place and period, and the latter is a decentered view in which
> we seek to transcend the here and now.
>
> (1991, 1)

Thus, being decentered (perhaps discomforted) in a commemorative landscape that allows access and agency, helps visitors connect the lessons of what is being commemorated with their own life experience and needs, leading to a deeper connection that may forge a sense of witnessing and a transformative attachment to the memorial place.

Second, and relatedly, a memorial landscape must be understood within a larger scalar context. I have argued that May 4 at Kent State, though commemorated, has had a minimal impact on our respect for free speech because the commemorative landscape lacks a direct connection to the shootings' greater significance; it does *not* go beyond the loss of life (Post 2016). But, why were those lives lost? The three memorials highlighted in this chapter have successfully connected the microcosm of the memorial landscape to larger issues of violence in society. Their designs encourage, perhaps require, reflection and the realization of life's temporality.

This dialectic of scale and connectivity also reflects the place quality of being both inward and outward. As David Seamon explains,

> place as defined by its inward and outward aspects requires consideration of
> the possible range of ways in which a place does or does not connect itself
> with and respond to the larger world of which it is a part.
>
> (Seamon 2014, 15)

By actively bringing the visitor's body in the memorial landscape and giving it agency, the visitor is able to personally connect their own life experience to the larger concerns of violence in society and why such incidents occur. This personalization of the tragedy enables the visitor to bear witness, make place, and learn from the tragedy, emotional tools that transform the individual and hopefully prevent such tragedies in the future.

Anne Buttimer sums up much of this connection between person and place, past and present, and life and death by saying,

> This tension between stability and change within rhythms of different scales, expressed by the body's relationship to the world, may be seen as prototype of the relationship between places and space, home and range in the human experience of world.
>
> (1976, 285)

Thus, recognizing that one is alive (stability, presence) while witnessing the deaths of others (change, absence) creates a unique relationship – attachment – between that person and the place where that recognition occurs. Commemorative landscapes that give agency to the visitor by using not only sight but also touch, sound, and even smell help produce a corporeal experience that enables visitors to develop a strong connection to place. This chapter highlights the role of the body in the commemorative landscape and discusses how a deeply emotive attachment to place and bearing witness to violence can transform our view of the world.

References

Agnew, J. A. 1987. *Place and Politics: The Geographical Mediation of State and Society*. Boston: Allen and Unwin.

Alderman, D. H. and J. F. J. Inwood. 2013. Landscapes of Memory and Socially Just Futures. In *A New Companion to Cultural Geography*, edited by N. Johnson, R. Schein, and J. Winders, 186–197. Malden, MA: Wiley-Blackwell.

Alderman, D. H., P. Kingsbury, and O. J. Dwyer. 2013. Reexamining the Montgomery Bus Boycott: Toward an Empathetic Pedagogy of the Civil Rights Movement. *The Professional Geographer* 65: 171–186.

Associated Press. 1999. KSU to Mark Where 4 Died. *Cleveland Plain-Dealer*, p. B1, 12 January.

Buttimer, A. 1976. Grasping the Dynamism of Lifeworld. *Annals of the Association of American Geographers* 66: 272–292.

Carter, P. L. 2015. Where Are the Enslaved? TripAdvisor and the Narrative Landscapes of Southern Plantation Museums. *Journal of Heritage Tourism* 11: 235–249.

Casey, E. S. 1996. How to Get from Space to Place in a Fairly Short Stretch of Time: Phenomenological Prolegomena. In *Senses of Place*, edited by S. Feld and K. H. Basso, 13–52. Santa Fe: School of American Research Press.

Connerton, P. 1989. *How Societies Remember*. Cambridge: University of Cambridge Press.

Cosgrove, D. 1984. *Social Formation and Symbolic Landscape*. Madison: University of Wisconsin Press.

Doss, E. 2010. *Memorial Mania: Public Feeling in America*. Chicago: University of Chicago Press.

Drozdzewski, D., S. De Nardi, and E. Waterton. 2016. *Memory, Place and Identity: Commemoration and Remembrance of War and Conflict*. London: Routledge.

Entrikin, N. J. 1991. *The Betweenness of Place: Towards a Geography of Modernity*. Baltimore: Johns Hopkins University Press.

Foote, K. 2003. *Shadowed Ground: America's Landscapes of Violence and Tragedy*, 2nd ed. Austin: University of Texas Press.

Foote, K. 2016. On the Edge of Memory: Uneasy Legacies of Dissent, Terror, and Violence in the American Landscape. *Social Science Quarterly* 97: 115–122.

Freundlich, B. 2012. Empty Chairs on Quad to Honor Lives Lost in Pan Am Bombings. *Daily Orange*. http://dailyorange.com/2012/10/empty-chairs-on-quad-to-honor-lives-lost-in-pan-am-bombings/. Posted: October 22. Last accessed 15 March 2016.

Giroux, H. A. 2000. Public Pedagogy as Cultural Politics: Stuart Hall and the "Crisis of Culture." *Cultural Studies* 14: 341–360.

Harvey, D. 1979. Monument and Myth. *Annals of the Association of American Geographers* 69: 362–381.

Johnson, N. C. 1994. Sculpting Heroic Histories: Celebrating the Centenary of the 1798 Rebellion in Ireland. *Transactions of the Institute of British Geographers* 19: 78–93.

Kilgannon, P. 2014. Syracuse Students Remember Pan Am Flight 103 and Its Victims. Newhouse Communications Center. https://nccnews.expressions.syr.edu/education/su-students-remember-pan-am-flight-103-and-its-victims/. Posted: October 21. Last accessed 15 March 2016.

Leib, J. 2002. Separate Times, Shared Spaces: Arthur Ashe, Monument Avenue and the Politics of Richmond, Virginia's Symbolic Landscape. *Cultural Geographies* 9: 286–312.

Leib, J. 2004. Robert E. Lee, "Race," Representation and Redevelopment along Richmond, Virginia's Canal Walk. *Southeastern Geographer* 44: 236–262.

Lewis, P. 1979. Defining a Sense of Place. *The Southern Quarterly* 17: 24–46.

Loewen, J. W. 1999. *Lies across America: What Our Historic Sites Get Wrong*. New York: The New Press.

Lopez, B. 1990. *The Rediscovery of North America*. New York: Vintage Books.

McFarlane, C. 2011. Assemblage and Critical Urbanism. *City* 15: 204–224.

Mitchell, D. 2000. *Cultural Geography: A Critical Introduction*. Oxford: Blackwell.

Mitchell, D. 2008. New Axioms for Reading the Landscape: Paying Attention to Political Economy and Social Justice. In *Political Economies of Landscape Change*, edited by J. L. Wescoat, Jr. and D. M. Johnson, 29–50. New York: Springer.

Post, C. W. 2012. Placing Memory and Heritage in the Geography Classroom. *Southeastern Geographer* 52: 351–354.

Post, C. W. 2015. Seeing the Past in the Present through Archives and the Landscape. In *Social Memory and Heritage Tourism Methodologies*, edited by A. Potter, A. Modlin, D. Butler, P. Carter, and S. Hanna, 189–209. London: Routledge.

Post, C. W. 2016. Beyond Kent State? May and Commemorating Violence in Public Space. *Geoforum* 76: 142–152.

Rentschler, C. A. 2004. Witnessing: US Citizenship and the Vicarious Experience of Suffering. *Media, Culture and Society* 26: 296–304.

Riaño-Alcalá, P. 2015. Emplaced Witnessing: Commemorative Practices Among the Wayuu in the Upper Guarjira. *Memory Studies* 8: 282–297.

Richardson, M. 2001. The Gift of Presence: The Act of Leaving Artifacts at Shrines, Memorials, and Other Tragedies. In *Textures of Place: Exploring Humanist*

Geographies, edited by P. C. Adams, S. D. Hoelscher, and K. E. Till, 257–272. Minneapolis: University of Minnesota Press.

Ryden, K. C. 1993. *Mapping the Invisible Landscape: Folklore, Writing, and the Sense of Place*. Iowa City: University of Iowa Press.

Sack, R. 1997. *Homo Geographicus: A Framework for Action, Awareness, and Moral Concern*. Baltimore: Johns Hopkins University Press.

Sandlin, J. A. 2010. *Handbook of Public Pedagogy: Education and Learning Beyond Schooling*. New York: Routledge.

Savage, K. 2009. *Monument Wars: Washington, D.C., the National Mall, and the Transformation of the Memorial Landscape*. Berkeley: University of California Press.

Schein, R. H. 1997. The Place of Landscape: Conceptual Framework for Interpreting an American Scene. *Annals of the Association of American Geographers* 87: 660–680.

Seamon, D. 2014. Place Attachment and Phenomenology: The Synergistic Dynamism of Place. In *Place Attachment: Advances in Theory, Methods, and Applications*, edited by L. C. Manzo and P. Devine-Wright, 11–22. London: Routledge.

Till, K. E. 2005. *The New Berlin: Memory, Politics, Place*. Minneapolis: University of Minnesota Press.

Till, K. E. 2012. Wounded Cities: Memory-Work and a Place-Based Ethics of Care. *Political Geography* 31: 3–14.

Till, K. E. and A. K. Kuusisto-Arponen. 2015. Towards Responsible Geographies of Memory: Complexities of Place and the Ethics of Remembering. *Erdkunde* 69: 291–306.

Tuan, Y.-F. 1977. *Space and Place: The Perspective of Experience*. Minneapolis: University of Minnesota Press.

Tyner, J., A. S. Kimsroy, and S. Sirik. 2015. Nature, Poetry, and Public Pedagogy: The Poetic Geographies of the Khmer Rouge. *Annals of the Association of American Geographers* 105: 1285–1299.

University Communications and Marketing. 4 May 1999. Markers To Be Dedicated September 8 at Kent State University. University Press Release. 2 September.

Wagner-Pacifici, R. and B. Schwartz. 1991. The Vietnam Veterans Memorial: Commemorating a Difficult Past. *American Journal of Sociology* 97: 376–420.

Wasserman, J. R. 1998. To Trace the Shifting Sands: Community, Ritual, and the Memorial Landscape. *Landscape Journal* 17: 42–61.

Waterton, E. and J. Dittmer. 2014. The Museum as Assemblage: Bringing Forth Affect at the Australian War Memorial. *Museum Management and Curatorship* 29: 122–139.

6 Exploring place attachment and the immigrant experience in comics and graphic novels

Shaun Tan's *The Arrival*

Steven M. Schnell

Place attachment among immigrants is a complicated topic. Immigrants invariably straddle multiple worlds, places, and identities. They experience first of all, a rupture with their old home. Whether fleeing something or moving towards something hypothetically better, or some combination of both, immigrants find themselves in new, alien places. They must then set out to make these new worlds their home. In the process, the experience of place becomes transformative.

A key part of this process is establishing place attachment – which, at its heart, involves identifying with a place as it comes to feel like home. That idea of belonging is central to the idea of place attachment – it is difficult to conceptualize of one becoming attached to place absent the development of a feeling of belonging in that place (Armstrong 2004). Key to creation of this sense of belonging are the connections people form with others; such connections inexorably become a part of the experience of, and attachment to, place (Smith and Cartlidge 2011; Smith and McAlister 2015).

The transitory nature of the migrant's place experience has led some to conclude that place attachment and mobility are opposite ends of the place experience spectrum. Yet, although time spent in place has often been shown to be the best predictor of place attachment (Lewicka 2011), the correlation between time of residence and place attachment is only moderate; many other factors can influence it. Place attachment can occur even in the absence of long-term connections and cultural alienness (Lewicka 2014). Even when one is physically removed from place, place attachment can be maintained in many ways, including family and social connections, transportation, and communications advances (Barcus and Brunn 2010), as well as through art, cultural practices, and music (Smith 2002).

Much place attachment research focuses on quantitative measures and correlation analysis, which have tended to look at snapshots in time – how attached to place someone is – rather than at change over time. As a result, most research tends to view place attachment as a static characteristic, rather than an evolving sense of self and belonging. By studying place attachment among immigrants, however, we can move past this static view and instead explore the *process* by which place attachments are created. This in turn adds to our understanding of the ways that such attachments transform the individuals experiencing them. After all, process is one of three areas in place attachment research that Maria Lewicka (2011) argues is in need of further attention.

The power of the personal

Much research on place and immigration has taken as a given the importance of ethnic and national identification, as well as other socio-economic classifications. Indeed, the experience of place among migrants differs greatly according to gender, class, and ethnicity (Kaplan and Chacko 2015). Analytically, social scientists often turn to such categorizations and give them primacy in our analyses of immigrant experiences (e.g., *Journal of Cultural Geography*'s February 2015 special issue on immigrant identity and place). The place attachment literature has likewise tended to focus on broad relationships between particular groups and quantifiable degrees of place attachment (Lewicka 2011). Setha Low has even argued that place attachment requires a symbolic relationship between a group and a place (Low 1992).

However, immigrants often also construct identities that do not rely on traditional categories of identity such as nation or ethnicity, resulting in identities that blur and blend such old distinctions (Ehrkamp 2005). Though categorical identities matter, so too do the local and the individual (Kaplan and Chacko 2015). Place attachment is multidimensional, and can change at different scales and different times, and among different individuals. It can draw on family and social networks, or can be more symbolic in nature. It can be at the individual or the group level (Barcus and Brunn 2010). To further understand the processes of building place attachment, therefore, individual case studies can provide insight that categorical statistical analyses cannot.

So how can scholars approach the more subjective and personal aspects of place attachment? One possibility is through examination of artistic creations. Helen Armstrong has argued that, in studying the relationship between migration and place, one must move beyond numeric data, and draw on sources as diverse as art and philosophy (2004, 238). John Tomaney has similarly stated that, "poets and novelists allow us imaginatively to inhabit local worlds through the use of tropes and narrative forms that plausibly thread together complex and particularistic interrelationships among space, place, and environment" (2012, 668). To these calls, I would add comics as another fruitful source for understanding the connections between people and place.

Comics, geography, and place

Comics and graphic novels,[1] previously dismissed as "kids' stuff," have begun to be treated as serious works of art and as cultural productions worthy of study. With the publication of Will Eisner's *A Contract with God* (1978) and Art Spiegelman's *Maus* (1980–91), it is clear to many that such forms of graphic storytelling provide a unique venue for exploring serious topics. But comics have only recently become the focus of geographical analysis (Dittmer 2014b). While text-based literature has been an ongoing topic of investigation by cultural geographers since the 1970s (e.g., Aiken 1977, 1979; Pocock 1988; Sharp 2000), comics have been relatively understudied.

Jason Dittmer, more than any other geographer, has brought comics studies into the field (e.g., 2005, 2007a, b, 2011, 2014a, b, c; Dittmer and Larsen 2007). Dittmer's most enduring focus has been on the dialogue between national identity and comics – the way that nationalist superheroes such as Captain America, Captain Britain, and Captain Canuck have both shaped and been shaped by senses of national identity and geopolitical discourses among their audiences. Others such as Oliver Dunnett (2009, 2014) have picked up Dittmer's focus on identity and geopolitics. Comics have also been used as a means of exploring a diverse range of geographical topics (Dittmer 2014b, c), including gender and racial representations (MacLeod 2014), the role of comics in creating distinctive local culture (Huston 2014), the use of comics in multi-media artistic performance in specific spaces (Venezia 2014; Dittmer and Latham 2015), the blending of local and outsider voices to examine conflict zones (Fall 2014; Holland 2014), and the ways that comics employ space to create narrative (Round 2014; Goodrum 2014; Doel 2014).

What few have examined is the subjective experience of place and place attachment as depicted in comics. In many ways, comics provide an ideal medium for portrayal of place. Traditional text narratives provide sensual input filtered through the lens of language (rendering the senses into word-based descriptions). Comics, on the other hand, also provide a non-language-based experience of scene and place, one encountered by the reader through direct visual stimulation, rather than place experience as filtered through linguistic filters and categories. Place depictions in this medium could thus be argued to have a more visceral effect on the reader. Comics are also an inherently spatial medium. Not only is there a narrative flow, there is also a flow of images through space (on the page, from page to page), and there is a dimensional quality to their rendering of place. Such characteristics allow the creators of comics to use words and images to create a unique sense of place, and to provide the reader with a potentially insightful window into the processes of place attachment.

So, how can images, words, and the interplay between them in graphic novels help the reader to understand the experience of place? What can we learn about place attachment, about alienation from place, and about becoming part of place (or failing to) from the work of some of the masters of the art form? This chapter is an initial exploration of their potential, focusing on Shaun Tan's *The Arrival* (2006), a tale of an immigrant's experience in a new land.

Reading textual description of visual images is, in the end, only a pale substitute for experiencing the original work. I encourage the reader to examine *The Arrival* in its entirety . . . go ahead, I'll wait! It, and most graphic novels and comics, gain their power not from individual images or elements such as those highlighted here, but rather from the experience of reading, and viewing, the collective whole. I hope that my analysis here helps to illuminate some of the ways we can learn about place, place attachment, and immigration from such works, and perhaps point to new ways for scholars to understand and articulate our relationships to place.

The Arrival

Shaun Tan's masterful book *The Arrival* (2006) is a graphic novel in the truest sense of the word – there are no words, at least no words you can understand. *The Arrival* focuses on the experiences of a nameless protagonist, a man who leaves his fictional homeland, and moves to a new country, as he comes to make this new land his home. Tan is the son of a Chinese-Malaysian immigrant to Australia and a third-generation Irish-English Australian who spent years researching immigrant experiences while working on the book (Tan 2010b).

The sepia-toned images in the book are evocative of old photographs, to place them in an unspecified distant past, and they are created in a photo-realistic style, despite their depictions of fantastical places. The book opens with a series of vignettes of small items in the man's home – an origami bird, a clock, a child's drawing, a family portrait – ordinary and mundane aspects of daily life in the old country. We see him with his wife and daughter, clearly about to part ways. We don't know exactly why he's leaving, only that the place is shadowed – a flash-back scene shows the family walking down the street, with strange, dark tentacles beginning to engulf the town. Tan's artwork evokes, in a way words alone can't effectively portray, the sense of threat and looming danger that would lead some-body to leave all that is familiar. We do not know the details, nor do we *need* to know the details. Tan is after a direct feeling rather than a detailed explanation; the tentacles stand in for a wide range of possible dark menaces.

After his journey and arrival in an Ellis-Island-like facility, our protagonist sets out finding his way in the new world. *The Arrival* effectively portrays the sense of dislocation, of discombobulation, that comes with being immersed in an unfamiliar place; the effect builds through the accumulated details of Tan's artwork. In many ways, it is the things that are close, but not quite the same as previous experiences, that can be the most disorienting about a new place: unknown musical instruments, strange timekeeping devices, bewildering maps, and mysterious plumbing (Figure 6.1). Tan's new world is littered with such details; any one of which is odd, but which in total create a feeling of dislocation and surreality.

Disorientation is heightened by the lack of ability to comprehend words, things that had been second nature just weeks before. The only language found in the book is a written language, created by Tan by cutting up familiar letters and putting them together in unrecognizable fashion (Figure 6.2) (Tan 2010b). This can make even earning a living difficult, as Tan shows us in a sequence involving his first job – hanging posters – where he inadvertently hangs each of them upside down.

Tan never portrays this new place particularly foreboding or hostile, merely as mysterious, and certainly lonely, at least at first. As Tan has said in an interview, he purposely pruned out scenes of anti-immigrant sentiment in the new country that he had written into earlier drafts of the work, leaving the scenes of intoler-ance for the old world. This creates, as Tan said, a picture of how things should and could be, not necessarily how they are (Verstappen 2007). Apart from skin color, there are no overt markers of any specific cultural or ethnic group of the protagonist.

Figure 6.1 The disorienting details.

From *THE ARRIVAL* by Shaun Tan. Copyright © 2006 by Shaun Tan. Reprinted by permission of Scholastic Inc.

Figure 6.2 Somewhat familiar, yet incomprehensible: language in the new world.

From *THE ARRIVAL* by Shaun Tan. Copyright © 2006 by Shaun Tan. Reprinted by permission of Scholastic Inc.

One of the ways Tan explores the act of becoming part of place is by masterfully switching back and forth between the micro-level details and the macro-scale big picture, as we explore the overwhelming landscapes of oddness. Tan regularly looks at the same scene at multiple scales, starting with the personal and small, and cinematically zooming out to take in the whole scene. For example, in one sequence, we start with his view of a framed portrait of his wife and daughter hanging on the wall of his apartment. We then back out the window to see the side of his apartment building, where each window houses its own set of stories in isolation.

Tan then zooms out yet again to take in the full scene, and the individuals fade into insignificance – they're simply small parts of an overwhelming whole (Figure 6.3). His large full-page or two-page spreads create a magnificent broader sense of the strange place (Figure 6.4). As Tan himself said in an interview,

> I just find myself strongly attracted, in an empathetic way, to images of isolated figures moving through vast, often confounding landscapes. My intellectual self would say that this is a metaphor for a basic existential condition: we all find ourselves in landscapes that we don't fully understand, even if they are familiar, that everything is philosophically challenging.
>
> (Mitchell 2009, n.p.)

Figure 6.3 The final scene in a several-page-long sequence zooming out from the Arrival's apartment, located in the building to the left of center.

From *THE ARRIVAL* by Shaun Tan. Copyright © 2006 by Shaun Tan. Reprinted by permission of Scholastic Inc.

Figure 6.4 Vast, often confounding landscapes . . .

From *THE ARRIVAL* by Shaun Tan. Copyright © 2006 by Shaun Tan. Reprinted by permission of Scholastic Inc.

Part of what makes his explorations so effective is the use of the graphic novel form. As Tan has put it, text is much more linear, and prescribes a certain sequence of movement by the reader (Verstappen 2007). Although given pages and frames are intended to be "read" in a particular order, within a given spread, particularly the large two-page spreads that magnificently set the scene, there is no given sequence of images – the exploration of the world is left to the reader. And these detail-rich scenes reward slow, repeated examination and exploration. This is one of the advantages of the graphic novel format in depicting place and experience of it; we explore the world in a way somewhat similar to the newcomer. As Tan has put it, "having no narrative facility, we learn about the new world according to what we pay attention to, and what we choose to invest objects with meaning" (Tan 2010a, 5).[2]

What happens to gradually make this place less bewildering and more welcoming, to give the protagonist a sense of belonging and attachment? First and foremost, it is contact and connection. The latter part of the book consists of a series of small moments that gradually make this new place more like home. The first happens when he obtains a companion animal (Figure 6.5). Many people in this new country seem to have them, and his – a sort of cross between a dog and a tadpole – just shows up in his apartment one day. This creature serves as a helper and companion; it is a pet, sort of, but something completely different as well, with no direct analogue in our understanding.

Figure 6.5 The Arrival meets his companion.

From *THE ARRIVAL* by Shaun Tan. Copyright © 2006 by Shaun Tan. Reprinted by permission of Scholastic Inc.

Over time, other people begin to show him how to navigate and interpret his new home, such as in a sequence where the protagonist is trying to find something to eat, but doesn't know what to make of the tentacled fruit he encounters. In another sequence, a woman teaches him how to navigate the transit system, which seems to consist of flying boats (Figure 6.6).

Over time, his connections with others deepen as his helpers trade stories with him of their own immigrant experiences, recounting the darkness in their homeland that led them to move, or flee. The woman who helps him navigate the flying boats escaped a life where she was forced to give up her education and work in the drudgery of a filthy industrial job. The man who introduced him to tentacled strawberries and other exotic foods recounts his tale of unspeakable, genocidal horrors – depicted in a scene where ominous giants clad in hazmat suits towering over the buildings vacuum up fleeing citizens into giant portable vacuum cleaners. An old man in the factory where he finally finds work recounts how flag-waving patriotism in his homeland gave way to the reality of brutal killing fields. Subsequent pages illustrate men marching off to war amongst celebratory parades, followed by scenes of the chaos of war, and finally, by mounds of corpses and skeletons. In these spreads, we viscerally experience the dehumanizing impact of economic, social, and political forces that are beyond our control.

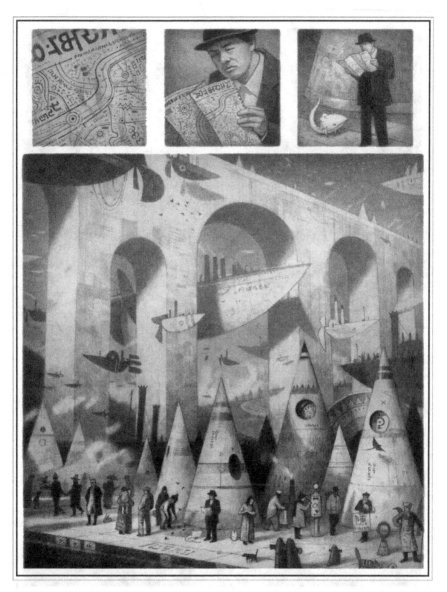

Figure 6.6 Finding the way in a bewildering new world; another example of Tan's zooming technique.

From *THE ARRIVAL* by Shaun Tan. Copyright © 2006 by Shaun Tan. Reprinted by permission of Scholastic Inc.

Figure 6.7 Dinner connections.

From *THE ARRIVAL* by Shaun Tan. Copyright © 2006 by Shaun Tan. Reprinted by permission of Scholastic Inc.

Our newcomer is then invited to dinner with a family, one whose structure mirrors his own back home (Figure 6.7). The families share stories, laughs, music, and gifts. Although the strange elements still permeate the scene – witness, for example, our host preparing dinner with a blowtorch – the domestic familiarity, hominess, and warmth permeate the scene. It is a scene that mirrors the protagonist's dinner table back home shown in the opening pages of the book (Figure 6.8). Similarly, he is introduced to playing a unique local game; the page-size landscape that follows this sequence glows with warmth and camaraderie.

As *The Arrival* comes to a conclusion, our protagonist is finally reunited with his family. In a scene that deliberately mirrors the first one of the book, as well as the scene where he visits his new friends for dinner, we see him at a meal with his family, giving instructions to his daughter to go out and obtain some food at the store. The strange place is now *home*.

Note here the parallels with the opening scene. Many of the same elements are found including the family photograph and the child's drawings. Notice how the origami animal, however, has changed from a bird to a . . . well, to a whatever-it-is-called and how the daughter's drawings now reflect the animals in her new home. The unfamiliar has become familiar, the foreign has become domestic, and the old has been merged with the new. Tan has referred to such parallels in his work as "rhymes," (Tan 2010a), that is to say, an element that visually resonates with and references something that went earlier, but that maintains its own distinctiveness. These rhymes, or parallels, are a key part of what allows our newcomer to begin to feel like his new place is *home* (Tan 2010b, 42). This is a key part of the building of place attachment in a new place.

Figure 6.8 A scene from the early pages back in the old country (left); the kitchen table in the new world once the Arrival's family has joined him (right). Note the parallels in composition as well as the integration of new world elements into the domestic scene.

From *THE ARRIVAL* by Shaun Tan. Copyright © 2006 by Shaun Tan. Reprinted by permission of Scholastic Inc.

The book concludes with his daughter heading out on her own into this new world. She navigates this new landscape fluently. In the street, she meets another new arrival, a young woman with her luggage, looking bewilderedly at her map. The closing page of *The Arrival* shows the protagonist's daughter pointing the new arrival in the right direction, providing the first step towards the new arrival herself becoming attached to this new place and making it her home.

The most powerful part of the work is the complete lack of (understandable) language. Helen Armstrong argues that "how people affiliate themselves and adapt to new situations involves processes which transform 'space' into 'place' by embedding culturally meaningful symbols into shared environments" (Armstrong 2004, 245). But for a newcomer, this is not possible. Part of what makes the migrant's experience so bewildering at first is the lack of understanding of various symbolism and meanings in the landscape. Such symbols abound in *The Arrival*, most notably the recurring image of the sunburst (e.g., Figures 6.1, 6.3, 6.4, 6.7, 6.8). As Tan himself has said about such symbols in his landscape,

> they don't stand for anything in particular. Even as their creator, I can't fully explain them, but they feel very important in this imaginary world. In a somewhat spiritual way, they offer a glimpse of the interconnectedness between all things, something that may be perceived without ever being fully understood.
>
> (Tan 2010b, 44)

They give the new land a distinct sense of place, and a conceptual unity, if one that is only vaguely understood to our newcomer.

The Arrival puts us, the readers, in the same shoes as the newcomer – simultaneously bewildered and enchanted, intimidated and coming to terms with this new place.

> It's hard to think of a time outside of childhood where the gap between object and meaning requires so much imaginative negotiation, in lieu of language, but the experience of immigration comes very close, and is an ideal subject for an illustrated story.
>
> (Tan 2010a, 4)

And yet, meaning does begin to emerge – incompletely, inconclusively, but enough to allow place attachment to begin to occur. Language is indisputably an important part of place-making and identity formation (Tuan 1991; Gilmartin and Migge 2015). What we see in *The Arrival*, however, is how much of the experience is pre-lingual – before one even has the words with which to describe one's experiences. Tan has said that the wordlessness also allows the reader to put their own interpretation onto the images, rather than having them dictated to them (Verstappen 2007), much as would be the case in the real world. As scholars writing in words, we tend to focus on language in our discussion of such topics. Yet, as David Seamon has argued, much of place's resonance is "pre-reflective" and unnoticed (2014, 14). This is part of the power of the graphic novel format; it breaks us out of the bias we have towards linguistic evidence, and gives us a new means for depicting human experience of place and the process of attachment to it.

The Arrival also allows us to experience what Seamon has called the "place ballet," that collection of individuals' relatively mundane daily routines that intersect in a particular place (Seamon 1980, 2014). Such routine interaction in place helps to lead to "'existential insideness', that is, a sense of belonging within the rhythm of life in place" (Lewicka 2011, 226). *The Arrival* repeatedly shows us small interactions in place, through which place attachment is deepened – as the protagonist goes about the basic day-to-day routine of living in place.

Another way that *The Arrival* effectively visually conveys the experience of making sense of a new place is through its destabilization of basic categories of understanding; this is central to the reader's experience of the book. Readers find themselves trying to make sense of what the pictures show us, trying to view the new world through the lenses and categories provided by our own experience. Those statues when the migrant enters the harbor are kind of like the Statue of Liberty. That floating boat, that's a transportation device – or at least, I'm pretty sure it is. But sometimes, there are no analogues, most notably in the case of the companion animals that many residents of the new country seem to have. Language fails us, and we have to just give in to the experience. Indeed, absence of language fluency, and even of linguistic translations of what is being observed, is central to the immigrant's experience of a new place, and Tan's work attempts to get across this sensual immersion in strangeness.

One other notable insight of *The Arrival* is Tan's focus on the protagonist as an individual, not as a member of a group. Group identity has often formed the focus of much research into immigrant identity and place, but Tan eschews any such framework of understanding; we know little of his background or culture, or even what sort of ethnic or national categorizations might exist in this world. The residents of his adopted new home hail from a broad range of backgrounds, as evidenced by their clothes and facial features. In the sequences where each of the protagonist's new friends details their reasons for migrating, Tan explores the horrors of war, genocide, and despotism. Much research has pointed to the importance of collective memory and knowledge of past history of place and family and the connections between them can give rise to a sense of place attachment (Tomaney 2015). Yet, in *The Arrival*, the bonding between the residents of the city stems not from their collective history in this new place, but rather from their common experience escaping terrible situations in their homes; the common bond to new place comes with its role as a place of refuge, safety, and possibility, and place attachment stems from this role. It is an idealized version of migration, by Tan's own telling:

> The unstated premise is that everyone in the city is actually an immigrant, each with their own stories of suffering and resilience. Collectively they have built a world inspired by idealism and the lessons of history; they are a community of people who know all too well what happens if tolerance, compassion and open-mindedness are allowed to fail.
>
> (Tan 2010b, 29)

When one is first coming to terms with place after migrating, Tan is arguing, one does not do it as a category or a group, but as an individual.

> The subject of immigration is often abstracted by the language of politics, economics and media, its themes generalized as a set of broad causes and consequences. In contrast, the anecdotal stories being told by immigrants themselves do not operate on the level of collective statistics at all. They are a constellation of intimate, human-sized aspirations and dilemmas: how to learn a phrase, where to catch a train, where to buy an item, whom to ask for help and, perhaps most importantly, how to feel about everything.
>
> (Tan 2010b, 10)

This early stage occurs prior to any broader landscape-building, group identity formation, or historical bonding – it occurs on a much more intimate and individualized level, focused on the mundane details of getting by. In *The Arrival*, place attachment begins through a series of small understandings and personal connections, and builds from there.

Learning from comics

So in the end, what can we learn about the processes of place attachment from *The Arrival*, a fictionalized tale rooted in an imaginary geography? First of all, we learn about the importance of the individual in the place attachment experience. Connection and attachment can only occur when the individual can overcome the dehumanizing effects of forces beyond their control and landscapes beyond their comprehension. Categorizations, such as often dominate studies of place attachment, can become prisons.

But even more importantly, graphic novels such as *The Arrival* greatly enrich our sensual experience of place, place experience, and place attachment. Much of what happens to our narrator is beyond words. The imagery of the book gives form gives voice and form to such feelings and sensations, and gets us closer to the preverbal sense of place as it is directly experienced. Whether through the sense of looming danger or hopeless servitude that arise in the old countries, the disorienting feeling of being immersed in this bewildering new world, or the growing sense of warmth and wonder that comes from increased connection and familiarity, *The Arrival* puts us more directly in the shoes of a new arrival.

The struggle to effectively integrate an identity forged in an old place with a new one lies at the heart of the book. Place attachment, the building of networks of human connection and meaning rooted in specific locations, is what allows us to survive in the face of larger-than-life, incomprehensible forces, and allows us to forge a human-scale existence in the face of an overwhelming landscape.

Today, in politics both in the United States and abroad, immigrants have been the subject of vociferous and vicious attacks, "othering" them as threats to a mythically pure culture. Many political leaders have done everything possible to alienate and turn people against immigrants, to demonize and scapegoat them, and to create a rigid line between a mythical "us" and a nefarious "them." What *The Arrival*

offers us, in contrast, is a deep understanding of humanity and place, and the connection between the two.

Notes

1 The distinction between comics and graphic novels is vague, and debates over the use of
such terminology have rarely provided much insight. Neither term is really an accurate
description of the vast array of graphic storytelling that has emerged. In the interest of
conciseness, I will use the terms "comics," "graphic novels," and "graphic narratives"
interchangeably here, with no implied difference in status, artistic merit, or form.
2 Education scholars have advocated using *The Arrival* for teaching English language
learners to navigate meaning without linguistic fluency (Farrell *et al.* 2010; Martínez-
Roldán and Newcomer 2011; Mathews 2014), and for addressing the experience of
immigration in the classroom (Boatright 2010).

References

Aiken, C. S. 1977. Faulkner's Yoknapatawpha County: Geographical Fact into Fiction.
Geographical Review 67: 1–21.

Aiken, C. S. 1979. Faulkner's Yoknapatawpha County: A Place in the American South.
Geographical Review 69: 331–348.

Armstrong, H. 2004. Making the Unfamiliar Familiar: Research Journeys towards Understanding Migration and Place. *Landscape Research* 29: 237–260.

Barcus, H. R. and S. D. Brunn. 2010. Place Elasticity: Exploring a New Conceptualization
of Mobility and Place Attachment in Rural America. *Geografiska Annaler: Series B,
Human Geography* 92: 281–295.

Boatright, M. D. 2010. Graphic Journeys: Graphic Novels' Representations of Immigrant
Experiences. *Journal of Adolescent and Adult Literacy* 53: 468–476.

Dittmer, J. 2005. Captain America's Empire: Reflections on Identity, Popular Culture, and
Post-9/11 Geopolitics. *Annals of the Association of American Geographers* 95: 626–643.

Dittmer, J. 2007a. The Tyranny of the Serial: Popular Geopolitics, the Nation, and Comic
Book Discourse. *Antipode* 39: 247–268.

Dittmer, J. 2007b. "America Is Safe While Its Boys and Girls Believe in Its Creeds!" Captain America and American Identity Prior to World War 2. *Environment and Planning D:
Society and Space* 25: 401–423.

Dittmer, J. 2011. Captain Britain and the Narration of Nation. *The Geographical Review*
101: 71–87.

Dittmer, J. (ed.). 2014a. *Comic Book Geographies*. Stuttgart: Franz Steiner Verlag.

Dittmer, J. 2014b. Introduction to Comic Book Geographies. In *Comic Book Geographies*,
edited by J. Dittmer, 15–24. Stuttgart: Franz Steiner Verlag.

Dittmer, J. 2014c. Comic Books. In *The Ashgate Companion to Media Geography*, edited
by P. C. Adams, J. Craine, and J. Dittmer, 69–83. Farnham, UK: Ashgate.

Dittmer, J. and S. Larsen. 2007. *Captain Canuck*, Audience Response, and the Project of
Canadian Nationalism. *Social & Cultural Geography* 8: 735–753.

Dittmer, J. and A. Latham. 2015. The Rut and the Gutter: Space and Time in Graphic Narrative. *Cultural Geographies* 22: 427–444.

Doel, M. A. 2014. Framing and So: Some Comic Theory Courtesy of Chris Ware and
Gilles Deleuze, Amongst Others: Or, an Explication of Why Comics Is Not a Sequential
Art. In *Comic Book Geographies*, edited by J. Dittmer, 161–180. Stuttgart: Franz
Steiner Verlag.

Dunnett, O. 2009. Identity and Geopolitics in Hergé's *Adventures of Tintin*. *Social & Cultural Geography* 10: 583–598.

Dunnett, O. 2014. Framing Landscape: *Dan Dare*, the *Eagle* and Post-War Culture in Britain. In *Comic Book Geographies*, edited by J. Dittmer, 27–40. Stuttgart: Franz Steiner Verlag.

Ehrkamp, P. 2005. Placing Identities: Transnational Practices and Local Attachments of Turkish Immigrants in Germany. *Journal of Ethnic and Migration Studies* 31: 345–364.

Fall, J. J. 2014. Put Your Body on the Line: Autobiographical Comics, Empathy and Plurivocality. In *Comic Book Geographies*, edited by J. Dittmer, 91–108. Stuttgart: Franz Steiner Verlag.

Farrell, M., E. Arizpe, and J. McAdam. 2010. Journeys across Visual Borders: Annotated Spreads of *The Arrival* by Shaun Tan as a Method for Understanding Pupuls' Creation of Meaning through Visual Images. *Australian Journal of Language and Literacy* 33: 19–210.

Gilmartin, M. and B. Migge. 2015. Home Stories: Immigrant Narratives of Place and Identity in Contemporary Ireland. *Journal of Cultural Geography* 32: 83–101.

Goodrum, M. 2014. The Body (Politic) in Pieces: Post 9/11 Marvel Superhero Narratives and Fragmentation. In *Comic Book Geographies*, edited by J. Dittmer, 141–160. Stuttgart: Franz Steiner Verlag.

Holland, E. C. 2014. Post-Modern Witness: Journalism and Representation in Joe Sacco's *Christmas with Karadzic*. In *Comic Book Geographies*, edited by J. Dittmer, 109–124. Stuttgart: Franz Steiner Verlag.

Huston, S. 2014. Live/Work: Portland, Oregon as a Place for Comics Creation. In *Comic Book Geographies*, edited by J. Dittmer, 59–71. Stuttgart: Franz Steiner Verlag.

Kaplan, D. H. and E. Chacko. 2015. Placing Immigrant Identities. *Journal of Cultural Geography* 32: 129–138.

Lewicka, M. 2011. Place Attachment: How Far Have We Come in the Last 40 Years? *Journal of Environmental Psychology* 31: 207–230.

Lewicka, M. 2014. In Search of Roots: Memory as Enabler of Place Attachment. In *Place Attachment: Advances in Theory, Methods and Applications*, edited by L. C. Manzo and P. Devine-Wright, 49–60. London: Routledge.

Low, S. M. 1992. Symbolic Ties That Bind: Place Attachment in the Plaza. In *Place Attachment*, edited by I. Altman and S. Low, 165–184. New York: Plenum Press.

MacLeod, C. 2014. From Wandering Women to Fixed Females: Relations of Gendered Movement through Post-Colonial Space in *Letters Doutremer* and *Le Bar du Vieux Français*. In *Comic Book Geographies*, edited by J. Dittmer, 75–90. Stuttgart: Franz Steiner Verlag.

Martínez-Roldán, C. M. and S. Newcomer. 2011. "Reading between the Pictures": Immigrant Students' Interpretations of *The Arrival*. *Language Arts* 88: 188–197.

Mathews, S. A. 2014. Reading without Words: Using *The Arrival* to Teach Visual Literacy with English Language Learners. *The Clearing House* 87: 64–68.

Mitchell, J. 2009. An Interview with Shaun Tan. *Bookslut*. www.bookslut.com/features/2009_07_014748.php.

Pocock, D. C. D. 1988. Geography and Literature. *Progress in Human Geography* 12: 87–102.

Round, J. 2014. Framing We Share Our Mother's Health: Temporality and the Gothic in Comic Book Landscapes. In *Comic Book Geographies*, edited by J. Dittmer, 127–140. Stuttgart: Franz Steiner Verlag.

Seamon, D. 1980. Body-Subject, Time-Space Routines, and Place-Ballets. In *The Human Experience of Space and Place*, edited by A. Buttimer and D. Seamon, 148–165. New York: St. Martins Press.

Seamon, D. 2014. Place Attachment and Phenomenology: The Synergistic Dynamism of Place. In *Place Attachment: Advances in Theory, Methods and Applications*, edited by L. C. Manzo and P. Devine-Wright, 11–22. London: Routledge.

Sharp, J. P. 2000. Towards a Critical Analysis of Fictive Geographies. *Area* 32: 327–334.

Smith, J. S. 2002. Rural Place Attachment in Hispano Urban Centers. *Geographical Review* 92: 432–451.

Smith, J. S. and M. R. Cartlidge. 2011. Place Attachment among Retirees in Greensburg, Kansas. *Geographical Review* 101: 536–555.

Smith, J. S. and J. M. McAlister. 2015. Understanding Place Attachment to the County in the American Great Plains. *Geographical Review* 105: 178–198.

Tan, S. 2006. *The Arrival*. New York: Arthur A. Levine Books.

Tan, S. 2010a. Words and Pictures, an Intimate Distance. Essay for ABC Radio National's "Lingua Franca" program. Transcript available at www.shauntan.net/images/essayLinguaFranca.pdf.

Tan, S. 2010b. *Sketches from a Nameless Land: The Art of the Arrival*. Sydney: Lothian.

Tomaney, J. 2012. Parochialism: A Defence. *Progress in Human Geography* 37: 658–672.

Tomaney, J. 2015. Region and Place II: Belonging. *Progress in Human Geography* 39: 507–516.

Tuan, Y.-F. 1991. Language and the Making of Place: A Narrative-Descriptive Approach. *Annals of the Association of American Geographers* 81: 684–696.

Venezia, T. 2014. 10th April, 1999, Conway Hall, Red Lion Square: *Snakes and Ladders*, Occult Cartography and Radical Nostalgia. In *Comic Book Geographies*, edited by J. Dittmer, 41–58. Stuttgart: Franz Steiner Verlag.

Verstappen, N. 2007. Shaun Tan [interview]. www.du9.org/en/entretien/shaun-tan922/.

Part IV

Restorative places

7 Constructing place attachment in Grand Teton National Park

Yolonda Youngs

Place attachment is a concept deeply rooted in geographic knowledge and theory. In his now classic books *Topophilia* (1974) and *Space and Place* (1977) geographer Yi-Fu Tuan encourages scholars to take a closer look at how we transform spaces of physical environments into places with deep and lasting meaning. As Tuan suggestions, we engage in a transformative process that creates places from spaces that carry the highly varied textures of emotional, social, and cultural meanings. We attach meaning through our actions, experiences, and memories. When it comes to America's national parks, this can be a highly personal or a collective social process. Each day thousands of visitors experience and enjoy the country's scenic treasures in a deeply individualistic and personal way. At the same time, national parks are places dedicated to a dual mission to protect the environment while providing recreational opportunities for the general public. National parks generate deep and lasting feelings of place attachment for many people. They are spaces at once intimately personal and private as well as public and shared.

While there are many important legal documents that provide a better understanding of the challenge and potential of national parks (Dilsaver 2016), there are two acts in particular that set the ground rules for national parks in the United States. These documents provide a key to understanding the distinct cultural, social, and political context of parks and how it translates into place attachment. On March 1, 1872 Congress created Yellowstone National Park "as a public park or pleasuring-ground for the benefit and enjoyment of the people" (Dilsaver 2016, 20). The *Yellowstone National Park Protection Act* created a precedent by setting aside federally owned and managed land for the specific purpose of public and national recreation. This legislation proved to be a powerful tool for natural and cultural resource protection. It spawned an international movement that now includes over 1,200 national parks in over 100 countries (USNPS 2016a). In the early 1900s, the U.S. government created other national parks and monuments but there was no unified, cohesive agency to manage this growing system. The War Department (U.S. Army) and the Department of Agriculture (U.S. Forest Service) filled this void to varying degrees of effectiveness, but in 1916 the tides turned.

The *Act to Establish a National Park Service* (Organic Act) of 1916 created the U.S. National Park Service (USNPS) as a division of the Department of the Interior and established the so-called *dual mission* of the agency. The mission requires that the agency 1) "conserve the scenery and natural and historic objects and wild life therein" and 2) "provide for the enjoyment of the same in such manner and by such means as will leave them unimpaired for the enjoyment of future generations" regulated for the "benefit and enjoyment of the people" (Dilsaver 2016, 34). In other words, the USNPS must provide for public enjoyment and access to the parks and monuments while also preserving and protecting their natural and cultural resources. This mission is an ongoing challenge and a balancing act for the modern-day agency. As of 2017, the USNPS is charged with preserving the considerable array of geographical, environment, and cultural resources in 417 units of the U.S. national park system spread across every state and territory. As for the "benefit and enjoyment" of the park formula, over 307 million people flocked to the parks in 2015, leaving the USNPS to find a balance between public enjoyment and environmental protection (USNPS 2016b).

Using Grand Teton National Park as a case study, this chapter explores the social construction of place attachment as both a personal and private experience that translates into a public and shared encounter through tourism and outdoor recreation. National park visitors are one part of this formula, seeking out public lands to restore and refresh themselves through hiking, camping, climbing, scenic driving, horseback riding, and other outdoor activities. But another, often overlooked, part of outdoor recreation and tourism are the seasonal employees and guides who work for park concession operators in national parks. They too seek out national parks as restorative places and often develop deep emotional bonds to specific parks. It is common for guides to return to the same park year after year or move to gateway towns during the off-season. After providing relevant background, this chapter dives in for a closer look at how place attachment plays out in the lives of pioneer river guides and concession operators in Grand Teton National Park. I suggest that a distinct cultural community of river runners and adventure outdoor recreationalists developed in Grand Teton National Park after World War II. This post-war era brought an outdoor recreation boom to much of the United States in a way that fostered an expansion of outdoor recreation. In Grand Teton National Park (GRTE),[1] a combination of environmental, cultural, and political forces combined with advances in outdoor recreation equipment and technology that shaped the community's evolution. Focusing on pioneer scenic river rafting guides and concession operators, I trace how these individuals developed a deep and lasting attachment to GRTE and how that connection, in turn, shaped a burgeoning commercial river rafting culture in the late twentieth-century western United States.

Restorative sanctuaries

William Wyckoff and Lary Dilsaver introduced the concept of restorative sanctuaries in *The Mountainous West: Explorations in Historical Geography* (1995). They proposed restorative sanctuary as one of five themes that define this region, noting that nineteenth-century Romantic ideas and the belief that mountains wielded special healing benefits fueled a special attraction to time spent in this region. Although the contemporary map of the U.S. national park system is not confined to the mountainous west, the movement to create these protected public lands can be traced back to this region. During the late nineteenth and early twentieth century, national park units were carved out of federal lands in the western United States in places such as Yellowstone, Yosemite, and Mount Rainer. All were noted for grand scenes of towering mountains, rivers, waterfalls, and forests. Kevin Blake's work on mountain symbolism expands on this theme by focusing on a variety of mountain ranges in Colorado (e.g., Blake 2002).

The concept of national parks as restorative sanctuaries attracts writers, journalists, and other scholars. Indeed, we can see threads of it woven into the writings of early national park proponents who worked to define the nascent national park idea and protect these public lands as the U.S. national park system took shape in the late nineteenth and early twentieth century. John Muir published eloquent essays that reinforce national parks as places of inspiration and health for an increasingly urbanized and industrial country. His 1901 book (*Our National Parks*) opens with:

> [t]housands of tired, nerve shaken, over civilized people are beginning to find out that going to the mountains is going home; that wilderness is a necessity; and that mountain parks are useful not only as fountains of timber and irrigating rivers, but as fountains of life.
>
> (Muir 1901, 1)

Later J. Horace McFarland, President of the American Civic Association, testified before Congress during the debates to create the U.S. National Park Service that the "parks are the Nation's pleasure grounds and the Nation's restoring places [*sic*]" (1916, 53). Decades later, as the national park system flourished, noted scholar and consultant to four directors of the USNPS, Freeman Tilden wrote that the "ultimate meaning of the national parks" is not that they are merely scenic places, or recreational spaces or tourist attractions, or the "special property of those who happen to live near them" (1951, 20–21). Instead, Tilden asserted, national parks are "really national museums . . . [t]heir purpose is to *preserve*" the natural and cultural resources for the public (22).

As part of the centennial anniversary of the USNPS in 2016, author and environmental activist Terry Tempest Williams published *The Hour of Land: A Personal Topography of America's Park*. She dives deeply into the emotional and psychological connection between people and national parks by weaving her signature style of personal recollections and family trips with broader social and cultural mediations. For Williams, national parks hold the power to transform personal place attachment into a movement that fosters public lands stewardship and a renewed agenda for social and cultural equity. "At a time when it feels like we are a nation divided, I am interested in how a sense of place can evolve toward an ethic of place, especially within our national parks" (Williams 2016, 14).

Williams also articulates the emotional and psychological effects of deep attachment to a place that is both personal and public space. National parks are deeply intimate spaces where one can spend an afternoon in solitude and quiet along high mountain trails or windswept plateaus. But that is far from the only or typical experience. Modern national parks of the twenty-first century are also places shaped by the forces of mass tourism. A visit to some of the more popular national parks during summer can involve jockeying with thousands of pleasure seekers for the perfect trail or optimal view, fording through traffic jams on crowded roadways, and navigating through tourist villages with hotels, restaurants, and shops. Williams meditates on the crossroads between personal/public and private/shared experiences in today's national parks. She writes,

> Acadia National Park is personal for me. If Virginia Wolf speaks of a room of one's own, how about a place of one's own, not to be shared or spoken of except with the 2.5 million other visitors that come to Acadia each summer? I am not alone in my affections. In 2014, Acadia ranked as one of the top ten most popular national parks in America.
>
> (Williams 2016, 89)

Then, in an echo of Muir's sentiments from more than a hundred years earlier, she writes

> For all of us, Acadia is another breathing space. Perhaps that is what parks are – breathing spaces for a society that increasingly holds its breath. Here on the edge of the continent in this marriage between wind and sea, the weaving of currents offers a tapestry of relief.
>
> (Williams 2016, 88)

Place attachment in national parks

People form deep, emotional ties places. Psychologists, sociologists, and anthropologists among other scholars have enthusiastically embraced the concept of place attachment (Manzo and Devine-Wright 2014). As a spatial and place-based discipline, geography traces a long history of ideas about the role of space, place, and process of place attachment in our everyday lives (Tuan 1974; Tuan 1977; Relph 1976; Masey 1997). Place attachment can engender a range of spatial scales, locations, and meanings. Jeffrey Smith and Jordan McAlister (2015) explore the deep emotional ties that rural Kansas residents feel for their homes as well as the county in which they live. Scholars from tourism and recreation studies also use the concept of place attachment but place an emphasis on understanding visitor experiences in national parks and tourist sites and the reactions of residents who live in gateway communities (Kyle *et al.* 2004; Gross and Brown 2008; Ramkissoon *et al.* 2012). Place attachment is gaining traction within a range of natural resource management fields as a way to unravel resource conflicts and stakeholder identification (Warzecha and Lime 2001).

David Harmon (2016) traces some of the many threads of emotional and psychological attachment that create a sense of place in national parks. While Harmon praises the superlative scenery and recreational qualities in national parks, he urges us to remember that national parks offer more than just pretty vistas. Here we see that the "power of place is not limited to nature. Historical and other cultural sites make up far more than half of the national park system. At some of these, we gain entry primarily through emotion" (2016, 23). For Harmon, battlefield sites (e.g., Minute Man National Historical Park) are particularly evocative landscapes that stir emotions connected to war, nationhood, and political upheaval. We could apply Harmon's appreciation of a sense of place in national parks to many locations. These places immerse us in a range of human emotions and challenge us to intellectually engage with different opinions, experiences, and perspectives than our own. Harmon writes that "[s]omething deeply fulfilling happens when we rise to meet these challenges, and this special kind of satisfaction can only be had in that park, that place, and nowhere else" (2016, 21). Each national park evokes a distinct emotional and psychological reaction grounded in that specific place while also connecting us to larger meanings and values that make up the tapestry of America.

Scholars from a wide range of fields including geography (Dilsaver and Colten 1992; Wyckoff and Dilsaver 1997; Colten and Dilsaver 2005), history (Rothman 1998; Runte 2010), anthropology, art history (Lippard 2000), and tourism studies are increasingly interested in the cultural and social history of tourist experiences in national parks. This research area continues to grow as more cultural and historical geographers find national parks to be complex and dynamic places to study a range of nature-society interactions. Topics include the value of national parks as sacred places and sources of national identity (Dilsaver 2009, 276), sites of contested social and cultural meanings (Morehouse 1996; Young 2002; Algeo 2004; Smith 2004; Vale 2005; Finney 2014), places of popular geographic imagination (Wyckoff and Dilsaver 1997), and iconic landscapes (Youngs 2012).

In GRTE some scholars suggest that place attachment could be a valuable natural resource management tool. Through a survey of visitor's attachments to the park places and specific management issues, scholars are learning that the park holds a "multitude of meanings for visitors, including emotional and social meanings" as well as those associated with scenic beauty and recreation (Smaldone *et al.* 2005, 90). But few studies have focused on the people who work and live, at least seasonally, in national parks and their feelings of place attachment. Each year, thousands of USNPS staff, volunteers, and concession employees live for several months in national parks around the country. They perform a wide range of essential duties and jobs to support park tourism, visitor services, and natural resource management. By looking at their stories, we can learn about the process of creating and maintaining place attachment for commercial employees in national parks.

River guides and their story

River guides are trained professionals who lead commercial rafting trips and frequently live seasonally in or near national parks with major whitewater and scenic river sections. They work for commercial businesses that hold a concession permit issued by the USNPS to offer tourist services in national parks. Depending on local conditions guides can work several short (2–4 hour) float trips per day or long, multi-day trips. The float trips along the Upper Snake River in GRTE are typically a few hours long. These short trips provide river guides with summers full of daily trips along the same stretch of river. This affords guides an opportunity to develop a deep sense of familiarity and intimacy with the river. Moreover, many park employees (including river guides) return year after year to the same park, in part, because of their deep and lasting emotional connection to the place. Due in large part to the type of work they do and the intimate connection they make with the local environment, many river guides become advocates who feel a "sense of ownership in local decision making," especially if it involves recreation and environmental management (Smith and McAlister 2015, 183). Different than other employees who might work inside at a visitor center or restaurant in the park, river guides work outdoors and are keenly aware of the park's environmental context. Changes in the river's flow, a new channel opening on the river, or even passing afternoon storms are all part of a river guide's daily life and work.

Part of a river guide's job also involves developing and communicating an accurate interpretation about the park's cultural and natural resources. A typical river trip includes short talks about local flora and fauna, river geomorphology, and the history of early land use and human settlement. These interpretive talks not only enrich a visitor's experience, but they also strengthen a river guide's emotional and psychological attachment to a specific national park. It is through their background research, the repeated telling of stories, and their daily experience of working on the river that river guides develop a strong attachment to place. By focusing on one set of national park employees (commercial river guides), I seek to add to our understanding of how place attachment develops in national parks. Some scholars propose using place attachment as a natural resource management tool in national parks (e.g., Smaldone *et al.* 2005), but few geographers have explicitly engaged place attachment theory to better understand how outdoor recreation guides form deep connections to national parks over time.

This research offers an innovative combination of theoretical approaches and methods to place attachment research. The methods for this research include a combination of archival research, oral history, and participant observation. Local and federal (USNPS) archives provide only a partial historical record of Upper Snake River rafting activity. Because this history and the GRTE archives that hold much of this material are relatively recent, the historical record is largely incomplete. Instead, the histories are stored in family archives of pioneer river guides and in their living memories. As a form of *participatory historical geography* (e.g., DeLyser 2014), I am actively collecting and preserving original historical records from pioneer guides and their shared family archives as well as recording oral history interviews from these guides. This chapter draws upon those oral histories and original family records. I recorded these histories through open-ended interviews. I am an environmental historical geographer as well as a former commercial river guide along the Upper Snake River. My identity makes this a participatory project in more ways than one as I tap personal networks in the river guide community and bring the perspective of a seasoned river guide to my research.

Oral history is an increasingly popular tool among geographers (Ward 2012; de Wit 2013). Ashley Ward suggests that oral histories provide "geographers the opportunity to examine the complexities and intricacies of place" (Ward 2012, 134). In her study of the people who worked in and near the tobacco factories of North Carolina, Ward encourages readers to consider the deeply intertwined aspects of people and place in stories that "lived in the individual memories of those that worked here, the collective memory is bound to this place" and it can be a "site of contested meaning that may have been created by the past, but is imposing itself on the present" (Ward 2012, 133–134). The same can be said about the river guides in GRTE who have worked on the same stretch of river and with a tight group of friends and fellow river guides for decades. As a former river guide on the Upper Snake River, I can attest to these enduring bonds to community and place that led me to this project.

This chapter helps advance the concept of *public cultural geographies* (see: Wyckoff 2014; Delyser 2016). My research design includes explicit collaboration between members of the river guide community, USNPS archivists and museum staff, student research assistants, and myself. As the lead scholar, I designed the overall research project and collected the bulk of the initial data. However, that is only the beginning. Each member of this interdisciplinary research team works collaboratively to amass relevant information and shape the research direction. Through ongoing two-way conversations between USNPS staff, the river guide community, and me, the scope and perspectives of the research are revised and refined as more information comes to light. For example, members of the river guide community and concession owners contribute photographs and documents from their family archives and share their experiences, knowledge, and memories of the river through oral histories. They also actively shape interviews by drawing upon their background and experience. This makes for a very dynamic process where the research team collaboratively shapes the outcomes and some of the final applied products such as museum displays, interpretive public talks, and new archival collections. The goal is to create a richer history of the river and GRTE and preserve this history for future researchers. This chapter presents the results of my first year of research in a long-term, four-year project. It is a story of one national park and how a group of people working as river guides construct deep and lasting attachments to place. In GRTE, pioneer river guides and park businesses started offering commercial trips in the late 1950s and early 1960s. They were at the leading edge of a nascent profession that would evolve into a flourishing, multi-million dollar outdoor recreation industry in parks and rivers across the United States.

Building place attachment on the Upper Snake River

The study area for this project is a 25-mile stretch of the Upper Snake River from Jackson Lake Dam to Moose, Wyoming in GRTE (Figure 7.1). It is an easy stretch of river to love. A typical afternoon on the Upper Snake garners stunning views of the Teton Mountains, bald eagle and moose sightings, and the soothing sounds of water flowing over glacially smoothed cobbles. This stretch of the river is a favorite among commercial river guides and private boaters as few rapids block downstream travel and Jackson Lake Dam provides a consistent flow of water. However, in the early spring, the river is dangerous. Hypothermic-inducing cold water rushes downstream, large trees fall into the river and jam channels; the river changes almost daily. During my interview with Frank Ewing, he recalled the daily challenge of running the river,

> and there were several occasions I remember when those of us who were on the river on a daily basis would have to deal first of all with the surprise plugging of some particular channel we were accustomed to going through.
>
> (Ewing 2015)

The view of the park from the river encompasses striking contrasts and beauty. The Snake River flows through the valley floor at 6,800 feet while the soaring, 40-mile long Teton mountain range lies ahead with many peaks over 12,000 feet. GRTE is part of the Greater Yellowstone Ecosystem and is a popular tourist destination located only 10 miles south of Yellowstone National Park. Today the Upper Snake River through GRTE offers a range of recreational boating including commercial scenic floating and fly fishing as well an increasing number of private boaters, kayakers, and canoeists. However, like many rivers in the United States, the Upper Snake River was a relatively untapped recreational resource in the 1950s and 1960s when many of pioneer guides began their careers.

There are excellent studies of the environmental history and geomorphology of the Snake River (Palmer 1991; Huser 2001; Marston *et al.* 2005) and GRTE (Righter 1982; Betts 1991; Sandborn 1993; Daugherty 1999; Righter 2014; Johnson 2016). The culture of river guiding in national parks is catching the attention of more scholars (Nash 1989; Teal 1994; Leavengood 1999; Westwood 1997; Sadler 2006), but many of these show a geographic bias for Grand Canyon National Park and the Colorado River; often overlooked is the parallel universe of boating and river guiding culture that developed on the Upper Snake River in the post-World War II era. Like many outdoor recreation activities such as skiing, hiking, and climbing, river rafting was not a new sport in 1950. However, technology that developed during World War II significantly changed river rafting in terms of boating gear, materials, and safety equipment. After the war, boats previously employed in combat became part of a large surplus of equipment. River guides in GRTE and other national parks repurposed these military surplus boats for recreation. In the process, they ushered in a new era of commercial river running that allowed more people to float on rivers in larger rubber rafts.

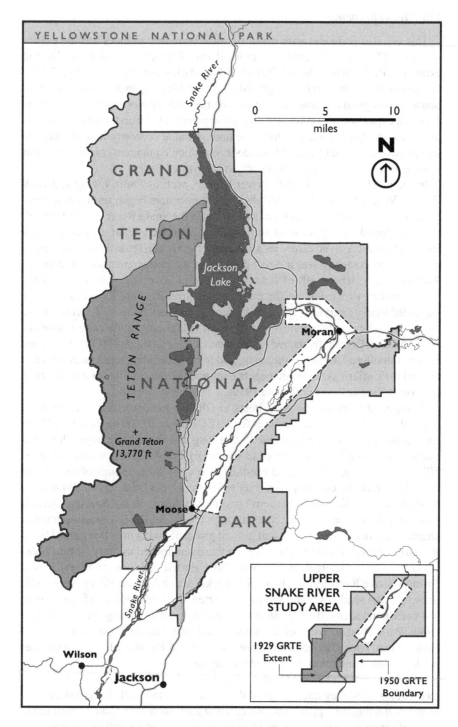

Figure 7.1 Map of Grand Teton National Park. The study area is a 25 mile stretch of the Upper Snake River from Jackson Lake Dam to Moose, WY (shown in the white boxed area). Note the original Park boundaries in 1929 and the expansion in 1950. Data: USNPS.

Cartography by Robert Edsall and author

This era also coincides with major political boundary changes and the expansion of GRTE. In 1950, the U.S. National Park Service increased the size and extent of GRTE. While the original 1929 boundaries encompassed primarily the Teton mountain range and a few glacial lakes, the 1950 expansion moved the park boundaries beyond the mountains to include the Snake River and its broad valley. The park expansion also coincided with an unprecedented recreation boom sweeping across the United States. More free time, a stable economy, and advances in inexpensive and widely available outdoor recreation equipment combined to spur a surge of tourists to national parks across the United States.

In the newly expanded GRTE, river runners such as Frank Ewing and Dick Barker along with local lodge and dude ranch operators began testing new ways to use rubber military surplus boats to float down the Snake River. As time passed, they developed strong personal and emotional connections to the place. Frank Ewing started his river career as a boatman for Grand Teton Lodge Company in 1957. The company was a direct product of the entrepreneurship of John D. Rockefeller, Jr. whose gift of land to the federal government fueled the 1950 park boundary expansion. As the park grew in size, Rockefeller saw a need for expanded tourism services. In 1955, he funded the construction of Jackson Lake Lodge and other visitor services in two small villages along the eastern shore of Jackson Lake. Eventually named Grand Teton Lodge Company (GTLC), the park concession offered fishing, sightseeing, and horseback riding tours. In an effort to expand their offerings, lodge managers explored the possibilities of offering boat trips on the Snake River in the late 1950s.

A shipment of rubber pontoon boats to GRTE's headquarters office inspired some of the earliest boating ventures for GTLC (Huser 2001; Hoops 2009). These rugged boats were between 20 and 30 feet long and shaped like large, elongated ovals. During the war, they served as temporary bridges. Their war-tested durability and size made them good candidates to carry large groups of tourists on the cold waters and cobble-paved riverbed of the Upper Snake River. With some adaptations including large "sweeps" (wooden rudders) on the bow and stern, it was possible to maneuver the boats through the Snake River's massive braided channel network. Ewing was one of a small group of young men that guided boats and passengers down the Snake River on these repurposed military surplus boats for GTLC. In 1963, Frank and his wife Patty started their own company (Frank Ewing's Snake River Float Trips). Dick Barker was a Jackson Valley native who grew up in the shadows of the Tetons. He started his river career as a fishing guide and occasional scenic float guide. As demand for scenic floating grew in the late 1960s, Barker and his wife Barbara joined Frank and Patty Ewing to create Barker-Ewing Float Trips. The company grew over time to include a large cast of river guides (women and men), boats, and touring vans.

Although Dick Barker and Frank Ewing started their businesses with small military surplus rafts, they later designed their own boats to handle the physical challenges of floating the Upper Snake River. Their ideas were influenced by their many years of running the river. Barker-Ewing's design proved so well adapted to local conditions on the Upper Snake River, that it became standard equipment for other river running companies in the area. Raft company brochures advertised a "Snake

Model" boat developed by Barker and Ewing and manufactured by Demaree Inflatable Boats, Inc. In 1985, Barker-Ewing Float Trips split into two separate family businesses with the Barker family operating the Upper Snake River scenic floats in the park while the Ewing family ran whitewater trips in the lower Snake River canyon south of the town of Jackson, Wyoming. Both companies continue to thrive and offer river trips to this day, albeit with some ownership and name changes.

But the boats were more than just material gear. They offered a medium for Ewing and Barker, as well as others in the valley, to experience the Upper Snake River in a way that few people did in the mid-twentieth century. These pioneer guides floated the same 26 miles of river most summer days from May to October and sometimes multiple times a day. They experienced the river and the valley's moods through spring hail storms, to summer rain showers and warm sun, to the first nip of fall's cooler temperatures. They saw a wide range of wildlife close up and personal every day as the rubber boats floated silently by the shoreline. Frank Ewing's distinct sense of place for the Snake River can be discerned even in his company's early tourist brochures through his eloquent writing. For example,

> [a] moderately early start will increase our chances of seeing wildlife along the river, however once we're underway the schedule is pretty much up to you. Occasional meanders ashore or down small side channels may reward us with a look at moose. . . . There's no button to push to make them just pop up, but these and many other animals . . . abound in the marshes along the way.
>
> (Ewing 1963)

Beyond boats and daily float trips, the people who worked for Barker-Ewing were fiercely loyal and shaped by the company's sense of environmental stewardship and dedication to accurate natural history interpretation about the park. Even though more thrilling whitewater and longer multiple week trips could be experienced on other rivers in the western U.S., many guides returned year after year to work for the Barker and Ewing families and row that same stretch of the Upper Snake River at the base of the Tetons. Interviews and company permit records attest that

> many of our former guides over the years have gone on to careers in related fields such as environmental mediation, fisheries and watershed management, wildlife photography, writing on environmental issues, etc. and they have told us that their time spent guiding here has had a direct influence on their ultimate choice of a career.
>
> (Barker-Ewing 1969)

Conclusion

This chapter explores the concept of restorative places, the idea of emotional bonding to national parks, and the feelings people experience while spending time in nature. Specifically, I trace how commercial river guides and commercial river company managers have developed a deep and lasting emotional connection to the upper Snake River and the environment within GRTE. Their place attachment may be seen in the careful attention to modified boat designs in tune with a specific location and stretch of river, their dedication to environmentally responsible business practices, accurate natural and cultural history interpretive tours for their guests, and the lasting ways this job influenced guides to pursue long-term careers related to environmental issues.

I suggest that a distinct cultural community of river guides and outdoor recreationalists developed in GRTE after World War II. This post-war era witnessed a boom in outdoor recreation throughout much of the United States in a way that fostered an expansion of hiking, camping, skiing, climbing, river rafting, and other types of recreation. In GRTE, a combination of physical, cultural, and technical forces shaped this community's evolution. These forces included technical and material changes in boats and gear over time, the challenges of navigating a dynamic and braided network of channels on the upper Snake River, and the close bonds between individuals that worked together each summer on the river. Individuals such as Frank Ewing and Dick Barker and their employee river guides developed a deep and lasting emotional connection to the Upper Snake River and GRTE. Their attachment expressed itself through the boats they designed in tune with their local river, their choice of environmentally related professions, and their status as stakeholders committed to stewardship of the Snake River and valley of GRTE through park management and town of Jackson decision making.

In the work and lives of these pioneer upper Snake River guides, we can better understand the transformation of space into place that Yi-Fu Tuan encourages his readers to consider. Through the lives of these guides we see the Upper Snake River shift from a space to a place that holds significant meaning. A first-year river guide spends their time learning the proper rowing techniques, the names of trees and birds, the names of the Teton mountain peaks, or the section of river that provides the best viewing angles of the mountains. But a seasoned guide such as Frank Ewing knows this river as a storied landscape full of memories, experiences, and emotion. "Occasional meanders or down small side channels" built Frank's vast knowledge of that river channel in every season and water level, offered an opportunity for a special moose sighting that Frank may still recall with fondness and awe, and creating a lasting memory that inspired him to write with precision in his own brochures.

National parks are indeed restorative places that inspire and reinvigorate the spirit of tourists as well as the people who work and live in these scenic locations. Anyone who has floated the Upper Snake River on a peaceful evening near sunset can attest to the healing, restorative powers of watching the sun fade in spectacular shades of pink and orange across the face of the Grand Teton or nearby peaks. The words of John Muir resonate as you realize that you may be one of those "tired, nerve shaken, and over civilized" people who could use more river time to reconnect with nature. Over time we become attached to these places because of the emotional benefits that we accrue. The dedication of Frank Ewing, Dick Barker, their families, and many pioneer river guides to the continued environmental stewardship of GRTE speaks to Terry Tempest Williams inquiry about the evolution of an "ethic of place." Frank Ewing was part of the community group that started the Snake River Fund, a grassroots organization dedicated to stewardship, public access, education, and partnerships for the river. Both the Ewings and the Barkers were instrumental in getting the U.S. government in 2009 to designate the Upper Snake River as a Wild and Scenic River. Ultimately, national parks are restorative to, and hold meanings for, both individuals and the public as they are both private and public places.

Note

1 GRTE is the official acronym for Grand Teton National Park as specified by the U.S. National Park Service.

References

Algeo, K. 2004. Mammoth Cave and the Making of Place. *Southeastern Geographer* 44: 24–47.

Barker-Ewing. 1969. Concession Report: US NPS Museum Collections and Archives. Grand Teton National Park. Administration Files. Folder C38. Barker-Ewing Scenic Tours, Inc., Contracts 1969, 2002–2003. Folder 2 of 4. Series 02.02.

Betts, R. 1991. *Along the Ramparts of the Tetons: The Saga of Jackson Hole, Wyoming.* Boulder: University Press of Colorado.

Blake, K. S. 2002. Colorado Fourteeners and the Nature of Place Identity. *Geographical Review* 92: 155–179.

Colten, C. and L. Dilsaver. 2005. The Hidden Landscape of Yosemite National Park. *Journal of Cultural Geography* 22: 27–50.

Daugherty, J. 1999. *A Place Called Jackson Hole: A Historic Resource Study of Grand Teton National Park.* Jackson, WY: Grand Teton Natural History Association.

DeLyser, D. 2014. Towards a Participatory Historical Geography: Archival Interventions, Volunteer Service, and Public Outreach in Research on Early Women Pilots. *Journal of Historical Geography* 46: 93–98.

DeLyser, D. 2016. Careful Work: Building Public Cultural Geographies. *Social and Cultural Geography* 17: 808–812.

de Wit, C. W. 2013. Interviewing for Sense of Place. *Journal of Cultural Geography* 30: 120–144.

Dilsaver, L. 2009. Research Perspectives on National Parks. *The Geographical Review* 99: 268–278.

Dilsaver, L. 2016. *America's National Park System: The Critical Documents.* Lanham, MD: Rowman & Littlefield Publishers.

Dilsaver, L. and C. Colten. 1992. *The American Environment : Interpretations of Past Geographies.* Lanham, MD: Rowman & Littlefield.

Ewing, F. 1963. *Snake River Float Trip: A Full Day of Safe and Quiet Adventure for You and Your Family in Jackson Hole Country.* Jackson, Wyoming: Self Published Brochure.

Ewing, F. 2015. Interview by Yolonda Youngs. Unpublished personal interview with Frank and Patty Ewing. Jackson, Wyoming. 22 June.

Finney, C. 2014. *Black Faces, White Spaces: Reimagining the Relationship of African Americans to the Great Outdoors.* Chapel Hill: University of North Carolina Press.

Gross, M. and G. Brown. 2008. An Empirical Structural Model of Tourists and Places: Progressing Involvement and Place Attachment into Tourism. *Tourism Management* 29: 1141–1151.

Harmon, D. 2016. Sense of Place. In *A Thinking Person's Guide to America's National Parks*, edited by R. Manning, R. Diamant, N. Mitchell, and D. Harmon, 20–29. New York: George Braziller Publishers.

Hoops, H. 2009. *The History of Rubber Boats and How They Saved Rivers*. Self Published. www.westwatercanyon.com/herm%20hoops/History-Rubber-Boats.pdf.

Huser, V. 2001. *Wyoming's Snake River: A River Guide's Chronicles*. Salt Lake City: University of Utah Press.

Johnson, N. W. 2016. *Shine Not in Reflected Glory: The Untold Story of Grand Teton National Park*. Morgan Hill, CA: Bookstand Publishing.

Kyle, G., A. Graefe, R. Manning, and J. Bacon. 2004. Effect of Activity Involvement and Place Attachment on Recreationists' Perceptions of Setting Density. *Journal of Leisure Research* 36: 209.

Leavengood, B. 1999. *Grand Canyon Women: Lives Shaped by Landscape*. Boulder: Pruett Publishing Company.

Lippard, L. 2000. *On the Beaten Track: Tourism, Art, and Place*. New York: The New Press.

Manzo, L. C. and P. Devine-Wright. 2014. *Place Attachment: Advances in Theory, Methods and Applications*. London: Routledge.

Marston, R., J. Mills, D. R. Wrazien, B. Bassett, and D. K. Splinter. 2005. Effects of Jackson Lake Dam on the Snake River and Its Floodplain, Grand Teton National Park, Wyoming, USA. *Geomorphology* 71: 79–98.

Masey, D. 1997. A Global Sense of Place. In *Reading Human Geography*, edited by T. Barnes and D. Gregory, 315–323. London: Arnold.

McFarland, J. H. 1916. Statement of Mr. J. Horace McFarland, President of the American Civic Association: National Park Service. *Hearing before the Committee on Public Lands, House of Representatives, Sixty-Fourth Congress, First Session, on H.R. 434 and H.R. 8668, Bills to Establish a National Park Service and for other Purposes*. 5 April and 6 April. Washington Government Printing Office, Washington, D.C.

Morehouse, B. 1996. *A Place Called Grand Canyon: Contested Geographies*. Tucson: The University of Arizona Press.

Muir, J. 1901. *Our National Parks*. New York: Houghton Mifflin Company.

Nash, R. 1989. *The Big Drops: Ten Legendary Rapids of the American West*. Boulder: Johnson Books – Big Earth Publishing.

Palmer, T. 1991. *The Snake River: Window to the West*. Washington, DC: Island Press.

Ramkissoon, H., B. Weiler, and L. D. G. Smith. 2012. Place Attachment and Pro-Environmental Behaviour in National Parks: The Development of a Conceptual Framework. *Journal of Sustainable Tourism* 20: 257–276.

Relph, E. 1976. *Place and Placelessness*. London: Pion Limited.

Righter, R. 1982. *Crucible for Conservation: The Struggle for Grand Teton National Park*. Jackson, WY: Grand Teton Natural History Association.

Righter, R. 2014. *Peaks, Politics, and Passion: Grand Teton National Park Comes of Age*. Jackson, WY: Grand Teton Natural History Association.

Rothman, H. 1998. *Devil's Bargains: Tourism in the Twentieth-Century American West*. Lawrence: University of Kansas Press.

Runte, A. 2010. *National Parks: The American Experience*. Lanham, MD: Taylor Trade Publishing.

Sadler, C. 2006. *There's This River . . . Grand Canyon Boatman Stories*. Flagstaff, AZ: This Earth Press.

Sandborn, M. 1993. *The Grand Tetons: The Story of Taming the Western Wilderness*. Moose, WY: Homestead Publishing.

Smaldone, D., C. Harris, N. Sanyal, and D. Lind. 2005. Place Attachment and Management of Critical Park Issues in Grand Teton National Park. *Journal of Park and Recreation Administration* 23: 90–114.

Smith, J. S. and J. M. McAlister. 2015. Understanding Place Attachment to the County in the American Great Plains. *Geographical Review* 105: 178–198.

Smith, L. 2004. The Contested Landscape of Early Yellowstone. *Journal of Cultural Geography* 22: 3–26.

Teal, L. 1994. *Breaking into the Current: Boatwomen of the Grand Canyon*. Tucson: University of Arizona Press.

Tilden, F. 1951. *The National Parks: What They Mean to You and Me*. New York: Alfred A. Knopf.

Tuan, Y.-F. 1974. *Topophilia: A Study of Environmental Perception, Attitudes and Values*. Englewood Cliffs, NJ: Prentice-Hall, Inc.

Tuan, Y.-F. 1977. *Space and Place: The Perspective of Experience*. Minneapolis: University of Minnesota Press.

U.S. National Park Service (USNPS). 2016a. *U.S. Department of the Interior: History*. www.nps.gov/aboutus/history.htm.

U.S. National Park Service (USNPS). 2016b. U.S. Department of the Interior. NPS Stats. Visitor Use Statistics. Grand Teton National Park. 1904 to 2016. https://irma.nps.gov/Stats/SSRSReports/Park%20Specific%20Reports/Annual%20Park%20Recreation%20Visitation%20(1904%20-%20Last%20Calendar%20Year)?Park=GRTE.

Vale, T. 2005. *The American Wilderness: Reflections on Nature Protection in the United States*. Charlottesville: University of Virginia Press.

Ward, A. R. 2012. Reclaiming Place through Remembrance: Using Oral Histories in Geographic Research. *Historical Geography* 40: 133–145.

Warzecha, C. A. and D. W. Lime. 2001. Place Attachment in Canyonlands National Park: Visitors' Assessment of Setting Attributes on the Colorado and Green Rivers. *Journal of Park & Recreation Administration* 19: 59–78.

Westwood, D. 1997. *Woman of the River: Georgie White Clark, White Water Pioneer*. Logan, UT: Utah State University Press.

Williams, T. T. 2016. *The Hour of Land: A Personal Topography of America's National Parks*. New York: Sarah Crichton Books.

Wyckoff, W. 2014. *Reading the American West: A Field Guide*. Seattle: University of Washington Press.

Wyckoff, W. and L. Dilsaver. 1995. *The Mountainous West: Explorations in Historical Geography*. Lincoln: University of Nebraska Press.

Wyckoff, W. and L. Dilsaver. 1997. Promotional Imagery of Glacier National Park. *Geographical Review* 87: 1–26.

Young, T. 2002. Virtue and Irony in a U.S. National Park. In *Theme Park Landscapes: Antecedents and Variations*, edited by T. Young and R. Riley, 157–182. Washington, DC: Dumbarton Oaks Research Library and Collection.

Youngs, Y. 2012. Editing Nature in Grand Canyon National Park Postcards. *Geographical Review* 102: 486–509.

8 Visitor perception, place attachment, and wilderness management in the Adirondack High Peaks

Tyra A. Olstad

[T]he view from where I sat is absolutely spectacular. From atop Algonquin Peak – the second highest mountain in New York State – I could look out and see dozens of other ranges and ridges, countless shimmering ponds and lakes, thousands of acres of thick, dark forest unfurling across a sea of state-protected wilderness. Although I knew there were, in fact, a few roads and towns tucked away within hiking distance and that the entire region was studded with lean-tos and laced with trails (affording "outstanding opportunities for a primitive and unconfined type of recreation," as per the Wilderness Act of 1964), the "imprint of man's work" was, per state and national definition, legislation, and enforcement, "substantially unnoticeable."

– Olstad (2015, 62)

When people speak of "*the Adirondacks*", they refer to a diversity of places and features in Upstate New York, ranging from rural towns and small cities to giant tracts of land owned by timber companies and other extractive industries. This precious place has thousands of freshwater lakes, tens of thousands of miles of fish-filled streams, rolling, bog-riddled lowlands, soaring, tundra-capped mountains, bears and moose, and the iconic loon with its haunting wail. Since the mid-nineteenth century, when people began visiting the Adirondacks in search of relaxation, recuperation, and/or adventure, and especially since 1902, when the State of New York drew the "Blue Line" expanding the borders of a "Forest Preserve" to create a State Park totalling 6.1 million acres, the Adirondacks have become one of America's oldest, best-known, and arguably best-loved wild places.

While many individuals express admiration for the Adirondack's lesser-known ponds, bogs, forests, and mountains, most people's attention centers on the "*High Peaks*" (46 summits historically believed to be above 4,000 feet in elevation). Predating the "*14er Club*" in Colorado by decades (see Blake 2002), the 46 peaks have long inspired hikers to earn the title of "*46R*" after checking off each summit. Thirty-six of these iconic landmarks are clustered in one management unit: the High Peaks Wilderness. With over 200,000 acres of mature forests, diverse habitats, glistening lakes, cascading streams, and ice- and water-sculpted valleys, the wilderness features the most remote, highest mountains in New York State. The High Peaks Wilderness is the largest, most-visited, and most iconic area within Adirondack Park. Each year, tens of thousands of people go to the High Peaks to hike, camp, and canoe. They relive memories, reach goals, and, overwhelmingly, experience a place promised (via Article XIV in the New York State Constitution, approved in 1894) to be "forever kept as wild."

What is "wild," though? While visitors' adventures may feel primitive and the landscapes they encounter may look primeval, the designated "wilderness" through which they travel is anything but pristine. Remnants of logging (the dominant land use before creation of the Adirondack Forest Preserve in 1885) remain in the form of old roads, rustic structures, and altered species assemblages. Even more prevalent are reminders of nineteenth- and early-twentieth-century recreation. Many of today's trails and campsites were cut, constructed, and popularized more than a century ago, when traveling through the mountains was truly a rough, adventuresome experience. Since the 1950s, when visitation rates began to skyrocket and development pressures intensified, and especially since the 1970s, when a newly-created Adirondack Park Agency (APA) issued an Adirondack Park State Land Master Plan (APSLMP) that officially defined and designated "Wilderness", land managers have had to develop and implement strategies to protect natural resources and minimize visitor impacts.

Unfortunately, the APA's mandates and the New York State Department of Environmental Conservation's (1999/2016) management plans and practices don't always align with the perceptions and desires of individual citizens. In an effort to preserve the very wildness that people remember and seek, agency officials have removed structures, rerouted trails, and increased regulations and enforcement. Although well-intentioned, many of these actions have been met with confusion, dismay, and heated debate, in part because the process of altering well-known landscapes and limiting familiar experiences threatens individuals' senses of place identity as well as their place-based aspirations and ideals.

Recent points of contention could be seen as just another chapter in what historian Philip Terrie refers to as a long legacy of "Contested Terrain" in the Adirondacks, but instead they have enabled the emergence of an ethic of "stewardship," in which individuals with a deep love for the High Peaks have voluntarily engaged in education, outreach, and boots-on-the-ground ecological restoration. This chapter uses qualitative content analysis of local media and online discussion forums (for methodology, see Henrich and Holmes 2013) as well as participant observation (Olstad 2015) to describe aspects of place attachment, visitor expectations, and land management priorities in the Eastern High Peaks Wilderness Area and discusses how an emerging stewardship-based approach is helping visitors engage in preservation and foster deeper emotional connections with the mountains.

What are the "Eastern High Peaks" and what is "Wilderness"?

Located within a day's drive of several major cities, including New York, Montreal, and Boston (Figure 8.1), Adirondack Park receives an estimated 10 million visitors each year (Adirondack Council 2013). Of the 140,000 hikers who seek out the High Peaks (NYSDEC 1999), an estimated 80 percent aim for the Eastern High Peaks (NYSDEC 1999), which include the highest, most accessible, and most iconic locations in the state – Mt. Marcy (the high point), Algonquin Peak (the

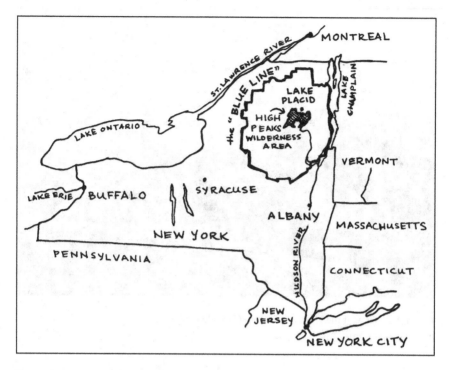

Figure 8.1 Location of Adirondack State Park.
Drawing by author

second highest mountain), and Marcy Dam (a popular camping spot and beloved scenic viewpoint) (Figure 8.2) . As described in the Unit Management Plan (UMP),

> The High Peaks region has helped define the Adirondack Wilderness [Figure 8.3]. For more than 150 years, it has been an immense attraction for those people with a sense of adventure and appreciation for wild places . . . [and] it has served an important role in introducing and educating people to the concept of wilderness.
>
> (NYSDEC 1999, 4)

"Wilderness," according to the APA, is the same in the Adirondacks as it is throughout America: "an area where the earth and its community of life are untrammeled by man – where man himself is a visitor who does not remain" (APA 2014, 19). Using the same language and standards as the federal Wilderness Act of 1964, the APA further defines wilderness as:

> an area of state land or water having a primeval character, without significant improvements or permanent human habitation, which is protected and managed so as to preserve, enhance and restore where necessary, its natural conditions, and which . . . has outstanding opportunities for solitude or a primitive and unconfined type of recreation.
>
> (APA 2014, 19)

Figure 8.2 View of the High Peaks from Cascade Peak.
Photo by author, 2015

Figure 8.3 View of the High Peaks from Heart Lake.
Photo by author, 2015

"Recreation" and "naturalness" can be somewhat contradictory, however. The UMP acknowledges, "One of the biggest challenges is how to keep the 'wildness' in wilderness and yet, still make it available for public use and enjoyment under today's heavy recreational pressures" (NYSDEC 1999, 2). State officials are beholden to stringently biocentric standards set by the APA in their Adirondack Master Plan, which state that

> protection and preservation of the natural resources of the state lands within the Park must be paramount. Human use and enjoyment of those lands should be permitted and encouraged, so long as the resources in their physical and biological context and their social or psychological aspects are not degraded.
>
> (NYSDEC 1999, 4)

In the Eastern High Peaks, skyrocketing visitation rates are making it nearly impossible to protect and preserve the resources. Usage at trailheads increased from 57,000 in 1983 to an estimated 140,000 in 1998 (NYSDEC 1999, 45), while the number of hikers reaching the main summits more than tripled from just over 10,000 in 2004 to more than 31,000 in 2015 (Rezin 2015). Park officials expect this trend to continue, forcing land managers to acknowledge and admit "high levels of resource degradation, and diminished opportunities for solitude, with the consequence that . . . wilderness standards are not being met" (NYSDEC 1999, 4).

Who are the visitors and what do they want?

Visitor demographics, desires, and experiences in the High Peaks Wilderness are fairly well documented. Numerous research projects have discussed topics such as: expectations and satisfaction (Dawson 2011, 2012), knowledge of cultural and natural history (Fredrickson 2002), resource substitutability (Graefe *et al.* 2010), displacement (Peden and Schuster 2009), stress and coping (Peden and Schuster 2004), perceived environmental impacts (van Riper *et al.* 2010), and the wilderness experience (Dawson 2006). While terms and numbers vary, surveys and interviews generally find that visitors perceive the Adirondacks in terms of exceptional beauty, serenity or peace, and wonderment (Fredrickson 2002). According to several studies conducted by Chad Dawson (2006, 2011, 2012), "the vast majority of people go to the mountains seeking to experience [the] natural environment and scenic beauty, and to enjoy the view from a mountain top; . . . Observe and hear wildlife in a natural setting; [and absorb] the tranquility & peacefulness of the remote environment. Crucially, these perceptions of "naturalness" emphasize a desire for an environment free and void of waste and human impacts" (Dawson 2012). People seek a pure, restorative experience in the High Peaks Wilderness (Wyckoff and Dilsaver 1995).

When asked to reflect on their trips, most visitors report feeling a deep connection to the wilderness as an important place in their life (Dawson 2012). This satisfaction seems to draw people back again and again, year after year, generation after generation. Of the hikers interviewed in the Northeastern Forest Preserve in

2011, an astonishing 99 percent were returnees (Dawson 2011). Even those who have not actually been to the High Peaks have already formed an impression of the wilderness as beautiful and serene based on information gleaned from friends and family, NYSDEC materials, tourism brochures and websites, media articles and reports, hiking organizations, and online forums that share information on Adirondack activities, history, and preservation.

Recreation research into visitor perceptions in the Adirondacks aligns with geographic research into sense of place and place attachment. Daniel Williams and Susan Stewart (1998) define sense of place as "the set of place meanings that are actively and continuously constructed and reconstructed within individual minds, shared cultures, and social practices" and "awareness of the cultural, historical, and spatial context within which meanings, values, and social interactions are formed" (19). This broad, fluid impression is, as Williams *et al.* (1992) elucidate, "often associated with an emotional or affective bond between an individual and a particular place . . . [which] may vary in intensity from immediate sensory delight to long-lasting and deeply rooted attachment" (31).

As Brian Eisenhauer *et al.* (2000) define, place attachment is an emotional bond between a person and a location "that is based on an appreciation for the land that goes beyond its use value" (423). Individuals or groups can form this relationship with particular sites, areas, or general physiographic types. In the case of wilderness, people can even form emotional and symbolic ties to the very idea of "wildness" (Williams *et al.* 1992). Margaret Mitchell *et al.* (1993) describe place attachment as entailing intimate knowledge of and respect for a location and a desire to continue to return to that place and/or share that place with others. For some people, place attachment can even deepen into *place identity*, in which a place becomes "a means of creating and maintaining one's self" (Williams *et al.* 1992, 32), or "cut[s] to sense of self" (Cheng *et al.* 2003, 96). When people fall in love with places, they see them not just as assemblages of biogeophysical properties (mountains, streams, lakes, etc.) or settings in which to engage in certain activities (hiking, camping, nature-gazing, etc.), but as memoried, storied, self-significant separate locations (Eisenhauer *et al.* 2000; Cheng *et al.* 2003; Chapin III and Knapp 2015).

Many visitors to the High Peaks Wilderness link their sense of self and mental state of being with wild mountains, waters, and woodlands (ADK Forum 2016). In iterations of the thread "Tell us about yourself" in the online discussion board *ADK Forum*, numerous discussants wrote variations of the following posts:

> Well, my love affair started with the 'dacks on the drive up. I couldn't get over the rugged beauty and the endless mountains and brooks, etc. . . . . We hiked Marcy on an overcast dreary day with absolutely no views, but I was hooked.
>
> ("adkdremn")

> [Climbing Tabletop] was the hardest thing that I have ever done . . . I was so banged up and bruised by the time we got down, but I wouldn't have changed it for all the tea in China. It was the most spiritually uplifting experience.
>
> ("Bristol")

Hiking together cemented [my] relationship [with my future wife]. . . . We talked about marriage during an aborted try up Dix one spring. The next year I officially proposed to her somewhere between Four Corners and Panther Gorge, on the way down from Skylight.

("Eagan")

Notably, in describing these events, each of these contributors names distinct geographic locations – Marcy, Tabletop, Four Corners (a trail junction) – as the unforgettable, irreplaceable settings for their *love affair, spiritually uplifting experience,* and *marriage proposal.* When individuals think of the High Peaks in terms of life-affirming events such as marriage (there were six proposals on Mt. Marcy in 2015) or scattering of a loved-one's ashes (an increasingly common occurrence), it shows how inextricably linked those spots are with their personal identity and sense of self-worth. Just as importantly, when individuals describe their time in the wilderness as healing and/or uplifting, they are, in fact, explaining their perception of it as a sanctuary – a place where they know they will find or recover beauty, and with it meaning and purpose. To paraphrase online contributor "Jackson," when places are imbued with personal significance, they become environmental repositories for and geographical manifestations of individuals' "own personal adventures, goals, memories, [our] own nirvana" ("Jackson," ADK Forum 2016).

As such, it's important to people that they be able to revisit spots that play an integral role in their aspirations or identity, or at least rest assured that the places remain as remembered, with their restorative qualities intact (Albrecht 2005, 2010).

What happens when visitors' sense of self-in-place and attachment are threatened?

Compared to areas that may face urban or industrial development, it would seem as though the High Peaks would, by virtue of their designation as "Wilderness", remain relatively unchanged and thus a safe haven for peoples' memories and dreams. Many people assume that designation alone assures preservation, but the NYSDEC counters:

Even [when the Adirondack Master Plan was adopted in 1972], it was obvious the High Peaks could not hope to accommodate large numbers of people without sustaining significant environmental and sociological change. Wilderness preservation could not exist without proper wilderness management; otherwise, wilderness designation would be meaningless.

(NYSDEC 1999, 86)

"Proper" wilderness management techniques are the subject of much debate. Some visitors and residents even argue that any type of deliberate "management" is antithetical to the ideal of natural and untrammeled wilderness (see ADK Forum 2016). With so many unique and sensitive ecological features, intense recreational use, and diverse experience levels, perceptions, and desires among visitors, the High Peaks have a legacy of complex management protocols. Typical of public land management issues nationwide, polarization can be attributed, in part, to sociopolitical factors. For example, Newman and Dawson (1998) describe how an adversarial "iron triangle" of public agencies, private interest groups, and the public "interlock[ed] to hinder action" during the 25-year-long UMP planning process. Of greater import to this chapter, James Cantrill and Susan Senecah (2001) found that late twentieth-century management decisions were "greeted by howls of protest" and dissolved into fierce ideological clashes not because of politics, ethics, or social identity, but due to a historically powerful, if ineffable, "Adirondacker sense of self-in-place" (195).

Unfortunately, the longer it takes to decide on and implement regulations, the more exposed places are to degradation (Newman and Dawson 1998). From the 1970s to the 1990s, public officials, non-profit organizations, and private citizens were increasingly dismayed to witness trails widening, campsites spreading, streams becoming non-potable, and mountaintop vegetation disappearing in the over-popular Eastern High Peaks. When the NYSDEC finally issued the UMP in 1999, they documented a long list of concerns, including but not limited to: crowded hiking and camping chokepoints, excessive wear and tear on campsites, numerous non-conforming structures, over-harvesting of campfire wood, trampling of rare and fragile alpine plants, and increased rates of soil erosion along trails and on summits (van Riper *et al.* 2010). With such a litany of threats, the restorative qualities of the place were at risk of vanishing.

In an attempt to rectify these issues, the UMP recommended several new regulations for the Eastern High Peaks (including group size limits, no open campfires, and no camping above 4,000 feet), approved the removal of some "non-wilderness-conforming" facilities, and proposed termination of "uses and activities that are not essential to wilderness management" (NYSDEC 1999, 89). Many of these decisions were met with resistance from visitors who associated their wilderness experience with such activities (Peden and Schuster 2004; ADK Forum 2016). Unsurprisingly, peoples' responses depend on their level of place attachment. Non-place-dependent visitors tend to be willing to shift their activities to other management units if they feel the regulations are prohibitively cumbersome (see "DSettahr," ADK Forum 2016). On the other hand, those who have a strong attachment to specific places must deal with the changes, largely by rescheduling trips, re-planning routes, and acquiring necessary equipment (Peden and Schuster 2004, 2009; Dawson 2012).

Interestingly, visitors with abiding attachment to the High Peaks and profound senses of place identity experience genuine ambivalence over management changes. They acknowledge feeling regret and loss (mainly due to an inability to relive or revisit cherished moments at favorite spots in the future), but also express appreciation for regulations meant to prevent further degradation and retain the restorative qualities of wilderness. For example, John Peden and Rudy Schuster (2009) found that "visitors with higher levels of experience use history were more sensitive to recreation impacts" (23) while, simultaneously, visitors such as "rdl" exclaimed, "the High Peaks are awesome. I've been going there for 30+ years and things have changed quite a bit in that time, primarily in the number of people who go, [but] the mountains themselves . . . are basically the same" (ADK Forum 2016).

Management debates take on a particularly personal tone when they involve site-specific alterations. After the NYSDEC surprised hikers by removing a few of the most-frequented (and most impactful) high-elevation lean-tos in the 1970s, many bemoaned the changes not in terms of economics or ecology, but in terms of individual memories and aspirations. For example, "Antler Perak" was upset by the removal of the shelter at Lake Tear, which he used to finish his 46 summits and thought of as "a gem", writing "Too bad it had to go!" (ADK Forum 2016). History was repeated when the UMP of 1999 called for the removal of several convenient, scenic, and memoried shelters en route to Mt. Haystack and Mt. Marcy. Visitors rued the losses. "I have hiked the Adirondacks for 10 years and I've always enjoyed the lean-tos," enthusiast Tom Faulkner was quoted in a *Washington Post* article on how the "rustic three-sided log structure that has sheltered Adirondack campers since the 19th century, is slowly disappearing from the landscape" (Hendricks 1996). "I remember some of the ones that have been removed," he insisted, "and I've missed them."

Lessons learned from Marcy Dam Pond: an example of place attachment, loss, and hope

Most recently, discussions about wilderness perception, management priorities, and place attachment flared up again in the public sphere due to the fate of a popular and beloved High Peaks landmark – Marcy Dam Pond (Figure 8.4). Although the

Figure 8.4 Photo of Marcy Dam Pond, slowly draining.
Photo by author, 2016

dam was technically an unnatural and "non-[wilderness]-conforming" feature, constructed in the 1930s by the Civilian Conservation Corps, it was still extremely popular with wilderness hikers, who loved the spectacular view of the pond that pooled behind it – classically serene mountains reflected in still water. Given its convenient location only a few flat miles from the trailhead, it was also an ideal place to set up camp and/or stage summit excursions. Countless visitors have memories like those of Adirondack-based writer Joann Sandone Reed, who recalls ski trips, runs, bike rides, and hikes to "the Dam" as an affirming and important part of her life (2015). As Phil Brown, editor of the online news journal *Adirondack Almanack*, explains, "Even people looking for wildness have come to see [human structures] like this as important landmarks . . . [They're] part of our history" (quoted in Mann 2015).

The NYSDEC had rehabilitated various features at the site, including degraded campsites, but had no real intention of altering the dam itself until 2011, when Hurricane Irene damaged the structure, allowing part of the pond to drain away and leaving the remnants in precarious danger of bursting. Officials were faced with the difficult decision of rebuilding the dam and restoring the pond, or removing it and allowing the pond to return to a natural marsh. Phil Brown's brief analysis – "hikers debated passionately whether the dam should be rebuilt to restore an iconic vista enjoyed by tens of thousands of visitors over the years" (Brown 2014). Such a statement barely begins to capture the vehemence with which people debated the situation. On one hand, people like "Bob" felt,

> [I]t's a Wilderness! . . . It's a place where by the DEC definition we are not supposed to feel the touch of man! . . . I am tired of these people who are outdoor advocates but have no real respect or concept of wilderness. [They want] to dam a natural waterway so [they] can have a nice view!?!
>
> (Brown 2014)

On the other hand, people like "Shawn Paradis" who possesses deep-seated emotional connections to the place felt, "I can only hope that Marcy Dam is restored to its original beauty" (Brown 2011).

In 2014, the NYSDEC announced plans to remove the structure, citing prohibitive costs, potential hazards, and conflicting management principles (Brown 2014). Debate flared up again. Some people, such as Neil Woodworth (executive director of the Adirondack Mountain Club), championed the decision, saying that "there is a growing number of people who appreciate nature reclaiming Marcy Dam and changing it into a different but still beautiful environment" (cited in Brown 2014). In essence, he was arguing that a new, even more wild and restorative place will emerge. Others, such as Joann Sandone Reed, expressed a combination of dismay, resignation, and nostalgia, saying "Marcy Dam had been a part of my life . . . Thanks for the memories Marcy Dam" (Reed 2015). As local journalist Brian Mann summarized, "It's not just the view, . . . people feel they're losing their personal history" (Mann 2015).

At least three lessons can be learned from the debate over Marcy Dam Pond. First, because "people's perceptions and evaluations of the environment are expressions of place-based self-identity" (Cheng *et al.* 2003, 96), management decisions that affect recreational experiences in well-known locations also affect individuals on visceral, personal levels. Environmental philosopher Glenn Albrecht recently went so far as to identify a condition he calls "*Solastalgia*" – a "psychoterric illness" that can arise when a person's well-being is negatively impacted by "loss of identity" associated with "loss of an endemic sense of place" (Albrecht 2010, 217–218). As he describes it, solastalgia involves:

> The pain or sickness caused by the ongoing loss of solace and the sense of desolation connected to the present state of one's home and territory. It is the 'lived experience' of negative environmental change manifest as an attack on one's sense of place. It is characteristically a chronic condition tied to the gradual erosion of the sense of belonging (identity) to a particular place and a feeling of distress (psychological desolation) about its transformation (loss of well-being).
>
> (Albrecht 2010, 226–227)

By this definition, individuals who are deeply attached to places like Marcy Dam Pond are, in essence, suffering from solastalgia.

Second, the NYSDEC typically justifies their management decisions based on political mandates, biocentric principles, and/or best-case scenarios, but when they implement their protocols and enforce regulations, they don't always communicate their rationale. This leaves citizens guessing as to the purpose (not to mention the effectiveness) of sometimes arbitrary-sounding changes to long-held practices and well-known places. Some visitors mistrust the agency, believing that:

> the DEC has a long record of not always doing what's best for the environment. They try, but often the politics and lack of manpower has them making compromises not always in the best interest of the wilderness they serve to manage (notice I did not say 'protect').
>
> ("Kevin," ADK Forum 2016)

Others chide land managers for failing to realize that "changing (or clarifying) regulations, while well intended, [is] useless unless accompanied by significant education for users to not just know, but understand" ("DuctTape," ADK Forum 2016). If officials better expressed their decisions in terms of ecological integrity and restoration, they might strengthen instead of threaten High Peaks visitors' attachment to the wilderness.

Third, and perhaps most interestingly, instead of simply feeling upset and helpless when it comes to general environmental degradation, increasing environmental regulation, and/or alteration of a beloved location, people are taking action. More and more outdoor enthusiasts who feel strongly attached to places within the High Peaks Wilderness are volunteering to help celebrate and protect the trails, summits, and scenic views they cherish and which give meaning to their life.

How is place attachment transforming into an ethic of place stewardship?

In the face of increasing visitation, decreasing opportunities for solitude, intensifying ecological impacts, and new and revised management protocols, those who are attached to the Eastern High Peaks have experienced one or several of the responses identified by Albrecht: some people feel a "sense of isolation about their inability to have a meaningful say and impact on the state of affairs that caused their distress"; others desire "to return to a past state/place where they felt more comfortable"; still others – including many in the Adirondacks – have "a strong desire to sustain those things [places] that provide solace" (Albrecht 2005, 44–45).

A desire to help sustain current natural conditions has become a motivating force for thousands of individuals who volunteer to act as stewards of the places that bring them life-affirming joy. *Soliphilia*, as Albrecht (2010) coined a term to capture the proactive side to solastalgia, is the "positive love of place, expressed as a fully committed politics and as a powerful ethos or way of life" (232). It seems natural, Laura Fredrickson (2002) agrees, "to suggest that when an individual develops a . . . special attachment to a particular place that the individual would extend a certain ethic of concern and care toward that particular place" (347). True to form, when Fredrickson queried High Peaks visitors "has your knowledge of the natural history of the Adirondacks stimulated an environmental ethic in you?," she found a "strong correlation between those . . . who felt a [strong attachment to] place and the likelihood of them possessing a preservation/environmental ethic" (Fredrickson 2002, 346).

On an organizational level, several groups now "provide path[s] for people to give back through stewardship" ("DSettahr," ADK Forum 2016). The ADK, for example, has expanded from its original emphasis on recreation and preservation to include education and stewardship. The club (now made up of 30,000 members) hosts programs for school groups, conducts Leave-No-Trace trainings, hires professional trails crews, and supports the "Summit Steward Program" (an initiative that aims to protect the rare and fragile alpine flora found atop only a few of the highest – and thus most visited and most trampled – peaks). As Summit Steward Coordinator Julie Goren said,

> There is no better way to protect these plants than by being up at the top where we can show people why it's important to stay on the rocks and enlist their help. These plants wouldn't exist on our Adirondack alpine summits today without hikers carefully choosing where to place their feet.
>
> (*Adirondack Almanack* 2015)

In the 26 years that the program has been running, it seems to have achieved the seemingly impossible goal of providing wilderness access while ensuring wilderness preservation. Stewards note steady improvement in the health and overall integrity of the mountaintop ecosystems, even while speaking with up to 500 hikers a day.

Similarly, the ADK 46Rs are not just a group of nearly 9,000 people who've sweated their way to all of the summits. Acknowledging that increasingly crowded trails are in danger of losing their physical and psychological wilderness qualities, the organization now emphasizes education and stewardship. In the spirit of Bob Marshall (the first person to identify and climb all 46 peaks over 4,000 feet), the ADK 46R's mission statement speaks not just about personal achievement, but about dedication "to environmental protection, to education for proper usage of wilderness areas, . . . and to the support of initiatives within the Adirondack High Peaks region by organizations with similar goals that enhance our objectives" (Adirondack 46ers n.d.). Their work is complemented by the ADK High Peaks Foundation, which funds "individuals or organizations whose activities provide a benefit to the New York State Forest Preserve and the people that use it", with "particular interest in funding wilderness zone and environmental protection, safe recreation, public education and biological research" (ADK High Peaks Foundation 2016).

Whether they participate in trail-maintenance days through the ADK 46Rs or contribute information to online forums, people's eagerness to help stems from their emotional connection to places and the ways by which those places resonate with their soul. For example, botanist Edwin Ketchledge established the summit steward program because he found great intellectual and spiritual satisfaction on the mountaintops and wanted to preserve and restore alpine ecosystems in the High Peaks. Likewise, Grace Hudowalski translated her grueling first hike up Mt. Marcy into an inspirational lifetime of hiking, outreach, and conservation:

> on the top just for a fraction of a moment, the clouds lifted while I was there and I looked down and there a mile below me was Lake Tear of the Clouds. . . . And you know, that did something to me. I had seen something – I felt it. I never forgot the mountain and I never forgot that trip.
>
> (quoted in Adirondack 46ers n.d.)

In 2003, a self-styled "simple group of forest preserve hikers and recreational users that cared deeply about the wild places where we choose to spend our time" and who, over time, "came to realize that it is up to us to help improve the public lands that we enjoy, making them better for those that follow us" (ADK High Peaks Foundation 2016) created internet communities that now host 5,000-plus members. Through their near-daily posts now numbering in the thousands, the group "look[s] for ways to improve and have a positive impact on the forest preserve areas we use." Countless individuals try to act on the sentiments expressed by "redhawk":

> [M]any of the things that I did years ago impacted the places I loved. But over time, with education, and even more importantly, seeing the same areas 20, 30 and 40 years later woke me up. So now I 'preach' about it . . . I think we should also be responsible enough to try to do what we can to preserve what we enjoy so much, for the future generations.
>
> ("redhawk," ADK Forum 2016)

Volunteer Summit Steward Jack Coleman agrees:

> Being a volunteer to me has never been about personal gain. It has always been more of what can I do to help . . . We all have a responsibility to share our experience and help protect the mountains we are so passionate about.
>
> (quoted in Bourjade 2013)

Moving forward

I end with observations made by F. Stuart Chapin III and Corrine Knapp (2015) who state: "places whose symbolic meanings are important to a person's identity . . . frequently motivate stewardship intentions" (41) and, in turn, "well-recognized actions that build place attachment could create a reservoir of potential stewardship" (38). Individuals, organizations, and land management agencies alike are dedicated to ensuring the Adirondacks are "forever kept as wild," but methods and motivations differ. APA representatives and NYSDEC officials are responsible for writing and enforcing regulations in the Eastern High Peaks, while recreation and conservation organizations are helping with outreach and communication as well as boots-on-the-ground trail maintenance and ecological restoration. In altering the recreational opportunities and changing the wilderness experience, even the most well-intentioned initiatives have the potential to threaten individuals' senses of place identity and their place-based aspirations and ideals. However, when management practices are aligned and implemented in terms of "soliphilia" and "stewardship," they can in fact draw upon and even strengthen place attachment. Love of place is a powerful motivating force, inspiring individuals to help care for the biogeophysical features and restorative qualities of the Eastern High Peaks. Practicing "stewardship", in turn, is an equally powerful way for visitors to deepen their relationships with the streams, ponds, forests, and mountains that bring such joy and meaning to their lives.

References

Adirondack 46ers. n.d. http://adk46er.org/about.html.

Adirondack Almanack Editorial Staff. 2015. High Peaks Summit Stewards Mark 400k Interactions. *Adirondack Almanack.* www.adirondackalmanack.com/2015/09/high-peaks-summit-stewards-mark-400k-interactions.html#comments.

Adirondack Council. 2013. State of the Park 2013: A Review of Elected and Appointed Government Officials' Actions Affecting the Adirondack Park. www.adirondackcouncil. org/uploads/sop_archive/1381887090_SOP2013.pdf.

Adirondack Park Agency. 1972 (revised 2014). Adirondack Park State Land Management Plan. http://apa.ny.gov/documents/Laws_Regs/APSLMP.pdf.

ADK Forum. 2016. www.adkforum.com/.

ADK High Peaks Foundation. 2016. Mission Statement. www.adkhighpeaksfoundation. org/adkhpf/main.html.

Albrecht, G. 2005. "Solastalgia": A New Concept in Human Health and Identity. *PAN (Philosophy, Activism, Nature)* 3: 41–55.

Albrecht, G. 2010. Solastalgia and the Creation of New Ways of Living. In *Nature and Culture: Rebuilding Lost Connections*, edited by S. Pilgrim and J. Pretty, 217–234. Washington, DC: Earthscan.

Blake, K. 2002. Colorado Fourteeners and the Nature of Place Identity. *The Geographical Review* 92: 155–179.

Bourjade, C. 21 November 2013. A Different Kind of Peak Bagger! *Today @ ADK* Blog of the Adirondack Mountain Club. www.adkmtnclub.blogspot.ca/2013/11/a-different-kind-of-peak-bagger.html.

Brown, P. 18 November 2011. No Decision on Marcy Dam. *Adirondack Explorer*. www.adirondackexplorer.org/outtakes/no-decision-on-marcy-dam.

Brown, P. 13 January 2014. DEC Plans To Dismantle Marcy Dam. *Adirondack Almanack*. www.adirondackalmanack.com/2014/01/dec-plans-dismantle-marcy-dam.html.

Cantrill, J. and S. Senecah. 2001. Using the "Sense of Self-in-Place" Construct in the Context of Environmental Policy-Making and Landscape Planning. *Environmental Science & Policy* 4: 185–203.

Chapin III, F. S. and C. N. Knapp. 2015. Sense of Place: A Process for Identifying and Negotiating Potentially Contested Visions of Sustainability. *Environmental Science & Policy* 53: 38–46.

Cheng, A., L. Kruger, and S. Daniels. 2003. "Place" as an Integrating Concept in Natural Resource Politics: Propositions for a Social Science Research Agenda. *Society and Natural Resources* 16: 87–104.

Dawson, C. 2006. Wilderness as a Place: Human Dimensions of the Wilderness Experience. *Proceedings of the 2006 Northeastern Recreation Research Symposium*. New York. GTR-NRS-14: 57–62.

Dawson, C. 2011. Northeastern Adirondack Forest Preserve Visitor Study. *Wilderness and Wildlands Research and Training*. www.esf.edu/nywild/publications/.

Dawson, C. 2012. Adirondack Forest Preserve Visitor Study. *Wilderness and Wildlands Research and Training*. www.esf.edu/nywild/publications/.

Eisenhauer, B., R. Krannich, and D. Blahna. 2000. Attachments to Special Places on Public Lands: An Analysis of Activities, Reason for Attachments, and Community Connections. *Society & Natural Resources* 13: 421–441.

Fredrickson, L. 2002. The Importance of Visitors' Knowledge of the Cultural and Natural History of the Adirondacks in Influencing Sense of Place in the High Peaks Region. *Proceedings of the 2001 Northeastern Recreation Research Symposium*: 346–355.

Graefe, D., C. Dawson, and R. Schuster. 2010. Roadside Camping on Forest Preserve Lands in the Adirondack Park: A Qualitative Exploration of Place Attachment and Resource Substitutability. *Proceedings of the 2010 Northeastern Recreation Research Symposium*: 205–213.

Hendricks, M. 14 September 1996. N.Y. Removing Lean-Tos from the Adirondacks. *The Washington Post*. www.washingtonpost.com/archive/realestate/1996/09/14/ny-removing-lean-tos-from-the-adirondacks/d5f6f3ad-e4a4-4987-a88a-5823b43ac9bc/.

Henrich, N. and B. Holmes. 2013. Web News Readers' Comments: Towards Developing a Methodology for Using On-Line Comments in Social Inquiry. *Journal of Media and Communication Studies* 5: 1–4.

Mann, B. 21 September 2015. Marcy Dam Dismantled, Altering an Adirondack Crossroads. *NCPR Regional News*. www.northcountrypublicradio.org/news/story/29560/20150921/marcy-dam-dismantled-altering-an-adirondack-crossroads.

Mitchell, M. Y., J. E. Force, M. S. Carroll, and W. J. McLaughlin. 1993. Forest Places of the Heart: Incorporating Special Places into Public Management. *Journal of Forestry* 91: 32–37.

Newman, P. and C. Dawson. 1998. The Interim Management Dilemma: The High Peaks Wilderness Planning Process from 1972 to 1997. *Personal, Societal, and Ecological Values of Wilderness: Sixth World Wilderness Congress Proceedings on Research, Management, and Allocation, Volume I.* USDA Forest Service Proceedings RMRS-P-4: 139–143.

New York State Department of Environmental Conservation. 1999. Wilderness Complex Unit Management Plan Wilderness Management for the High Peaks of the Adirondack Park. www.dec.ny.gov/docs/lands_forests_pdf/hpwump.pdf.

Olstad, T. 2015. Of Mountains and a Moth. *The Trumpeter: Journal of Ecosophy* 31: 61–69.

Peden, J. and R. Schuster. 2004. Stress and Coping in the High Peaks Wilderness: An Exploratory Assessment of Visitor Experiences. *Proceedings of the 2004 Northeastern Recreation Research Symposium*: 29–38.

Peden, J. and R. Schuster. 2009. Displacement in Wilderness Environments: A Comparative Analysis. *International Journal of Wilderness* 15: 23–29.

Reed, J. 20 September 2015. Memories of Marcy Dam. *North Country Public Radio.* http://blogs.northcountrypublicradio.org/allin/2015/09/20/memories-of-marcy-dam/.

Rezin, R. 2015. High Peaks Summit Stewards by the Numbers. *Adirondack Mountain Club Trailhead.* www.adktrailhead.org/high-peaks-summit-stewards-by-the-numbers-infographic/.

van Riper, C., R. Manning, and N. Reigner. 2010. Perceived Impacts of Outdoor Recreation on the Summit of Cascade Mountain, New York. *Adirondack Journal of Environmental Studies* 16. www.ajes.org/v16/vanriper2010.php.

Williams, D., M. Patterson, J. Roggenbuck, and A. Watson. 1992. Beyond the Commodity Metaphor: Examining Emotional and Symbolic Attachment to Place. *Leisure Sciences* 14: 29–46.

Williams, D. and S. Stewart. 1998. Sense of Place: An Elusive Concept That Is Finding a Home in Ecosystem Management. *Forest Science* 95: 18–23.

Wyckoff, W. and L. Dilsaver. 1995. *The Mountainous West: Explorations in Historical Geography.* Lincoln: University of Nebraska Press.

Part V

Validating places

9 Baseball stadiums and urban reimaging in St. Louis

Shaping place and placelessness

Douglas A. Hurt

Sport has the ability to unify disparate groups of people as well as to promote a sense of community and civic pride (Stone 1981; Raitz 1995a; Schimmel 2006; Zelinsky 2011). Sports landscapes are playing an increasing role in urban development in the United States, remaking cities as economic activities like manufacturing have declined in recent decades (Austrian and Rosentraub 2002; Turner 2002; Turner and Rosentraub 2002; Friedman *et al.* 2004; Hall 2004; Duquette and Mason 2008; Friedman 2010; Rosentraub 2014). Reconstructing the built environment to reallocate space to tourism, entertainment, and cultural spectacle is designed to fill this economic void and reshape urban identity. The process of creating new landmarks and identities is urban imagineering (or reimaging), the promotion of new urban place images (Archer 1997; Short 1999; Smith 2005). Entertainment zones, themed restaurants, atrium hotels, redeveloped waterfronts, and gentrified housing often accompany new sports facilities in American downtowns.

Stadiums are iconic urban spaces in the geographic patterns of sport. According to geographer Christopher Gaffney, stadiums are "sites and symbols of power, identity, and meaning that allows us to enter and interpret the cultural landscape through a common medium" (2008, 24). Others have couched the cultural significance of stadiums in religious terms calling them "among the few sacred grounds of a secularized society" (Neilson 1995, 30). They have even become tourist destinations (Gammon 2004; Ramshaw and Gammon 2005). Stadiums, and professional sport in general, are important shapers of urban identities and are woven into the social and economic fabric of American cities and their surrounding regions.

Since the 1870s, three permanent stadiums have hosted professional baseball games in St. Louis (Figure 9.1). The city's first baseball ballpark (Sportsman's Park, later renamed Busch Stadium) was located in a vibrant, working-class neighborhood. The second facility (Busch Stadium II) was constructed in the 1960s as part of a large urban renewal project on the southern edge of the urban core. The current stadium (Busch Stadium III) is located adjacent to the site of second ballpark but is the anchor of a planned mixed-use, baseball-themed entertainment district. This chapter examines the role that baseball stadiums in St. Louis, Missouri have played in both urban identity and peoples' emotional connection to place.

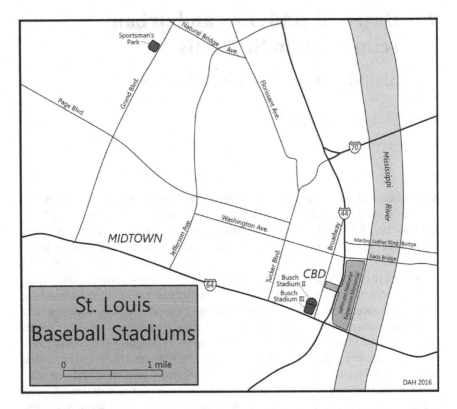

Figure 9.1 Map of the St. Louis urban core and Sportsman's Park, Busch Stadium II, and Busch Stadium III.

Cartography by author

Stadiums, sports landscapes, and the nostalgic past

Traditionally, the geography of sport has been an overlooked and understudied subdiscipline of geography (Bale 1988, 2003). As Peirce Lewis noted in the 1980s, "we need to know much more about the way that Americans have subdivided and allocated geographic place for recreation and games" (1983, 255). Due to their status as recognizable landmarks, sports stadiums have attracted the attention of geographers and a growing number are investigating the cultural significance of sports landscapes, their impact on city morphology, as well as the cultural spectacle surrounding sporting events.

For many urban residents, stadiums are sources "of civic pride and a symbol of victory and accomplishment" (Raitz 1995b, 5). Some people have compared stadiums to such iconic structures as Catholic cathedrals (Novak 1976; John 2002; Trumpbour 2007), Greek Agoras, and Roman Forums (Jackson 1984). Sports stadiums are gathering places that hold memories and help form the collective local identity of a city, region, and nation (Bale 1988).

Christopher Gaffney utilized case studies of Rio de Janeiro and Buenos Aires to investigate how stadiums are sites of power and identity in urban landscapes (2008). Although he focuses on soccer facilities, the ability of sport to transform the cultural landscape and shape cultural identities transcends specific sports. Gaffney's interpretation of stadium landscapes symbolization and the intersection of soccer with the maintenance and formation of racial, class, and national identities contributes to the larger discussion of the cultural impact of sport.

In his work on American college towns, Blake Gumprecht designated sports stadiums as a distinguishing characteristic of these unique urban places (2008). His case study of Auburn, Alabama investigated the historical evolution of athletics, particularly football, at Auburn University and the economic and cultural impact of games upon the host community. The home football game spectacle, Gumprecht concludes, is a key component of place-making and cultural identity for college towns.

Margaret Gripshover detailed the historical morphology of Wrigleyville, the North Side Chicago neighborhood surrounding Wrigley Field, the home of Major League Baseball's Chicago Cubs (2008). Predating the stadium, Wrigleyville was transformed from an ethnic enclave housing Germans and Swedes to a congested commercial, residential, and entertainment district to a modern mixed-use zone featuring entertainment, retail space, services, and residences. Real estate agents have successfully marketed the area as a desirable neighborhood for young, wealthy professionals (Gripshover 2008). LoDo in Denver, Colorado has followed a similar trajectory.

Finally, many investigations of sports landscapes involve how Americans remember and interpret the past. In general, nostalgia for historic times and lost places has shaped the collective memory expressed in a growing number of heritage displays in American public space (Lowenthal 1998; Hoelscher 2006; Doss 2010). In addition to the national proliferation of museums, monuments, memorials, and historic parks, since the 1990s the construction of new baseball stadiums has consistently utilized a retro aesthetic that infuses architectural elements of pre-1960 sports architecture onto modern facilities. Not surprisingly, nostalgic feelings shape how many Americans perceive their heritage, invest emotions in their surroundings, and foster emotional loyalty to teams and sports landscapes (Bale 1996; Ramshaw and Gammon 2005; Gordon 2013).

Using historic photographs and postcards, government reports and planning documents, as well as a synthesis of secondary source material, this chapter explores the historical geography of professional baseball stadiums in St. Louis and assesses their roles in fostering (or hindering) place attachment and in shaping the changing urban images of the city.

Baseball stadiums in St. Louis

Sportsman's Park (Busch Stadium) (1874–1966)

Sportsman's Park was home to the St. Louis Cardinals of the National League (beginning in 1920) and the St. Louis Browns of the American League (from 1902 until their move to Baltimore after the 1953 season). Renamed Busch Stadium in

1953 (but continually referred to as Sportsman's Park in the local vernacular) after Anheuser-Busch bought the stadium and Cardinals, the facility was demolished in 1966 and the site became home to a Boys and Girls Club (Reidenbaugh 1987).

Situated at the northwest corner of Grand Boulevard and Dodier Street in north St. Louis City, baseball was played at the 9-acre site on the edge of St. Louis as early as the 1860s (Reidenbaugh 1987). Like other ballfields of the time, Sportsman's Park was far removed from the valuable property of the compact urban core, located in the northwest city where land was readily available and inexpensive (Neilson 1986; Ritzer and Stillman 2001). Originally called the Grand Avenue Ball Grounds, the location was renamed Sportsman's Park in 1874 after the first wooden grandstand was constructed behind home plate (Reidenbaugh 1987). Frequent fires forced Sportsman's Park to be continually rebuilt until 1909 when the St. Louis Browns ownership located a new baseball diamond at the site, replaced the wooden structures with concrete and steel construction, added a second deck of stands from first to third base, built outfield bleachers, and raised the seating capacity to more than 18,000 fans (Reidenbaugh 1987). The improvements at Sportsman's Park were part of a national effort by baseball clubs to construct privately funded, durable facilities. Between 1902 and 1923, 14 (of 16 major league teams built larger stadiums with steel and concrete, creating the classic era of baseball parks (Neilson 1986; Ritzer and Stillman 2001).

In St. Louis, a permanent structure was fitting for an optimistic, growing urban area that had more than 575,000 people in the 1900 U.S. Census, the fourth largest city in the United States (Primm 1998; Sandweiss 2001). Population growth continued into the 1920s, as readily available manufacturing jobs attracted immigrants from Europe as well as rural-to-urban migrants from Missouri and the Mississippi River Valley (Primm 1998; Sandweiss 2001). City boosters confidently predicted continued population, infrastructure, and economic expansion as St. Louis planned to enlarge parks and public space in order to maintain its position in the top tier of American cities (The Civic League of St. Louis 1907).

As Sportsman's Park was further improved over time (including completing the double-deck grandstands, building a roof over the right-field pavilion, and adding arc lights for night games in 1940), it was constrained by the square city block that it occupied (Figure 9.2). By the 1940s the boxy facility had a capacity of 34,000 spectators, many of whom relied on the Grand Avenue streetcar line to reach games. A wing of double-decked stands extended along each foul line, circling behind home plate. The top rows of seats were covered by a narrow roof. A slender section of outfield bleachers ran from foul line to foul line. The field and bleachers were separated by an exterior brick façade from the sidewalks and streets outside. From the grandstands, patrons could look past the outfield beer advertisements and see linear streets with on-street parallel parking, rows of densely-packed two-story red brick row residential housing, family-owned business, commercial buildings, and occasional church spires (Bradley 2013). In sum, Sportsman's Park was a typical early twentieth-century ballpark, woven into the surrounding social and economic fabric of working-class German and Central European neighborhoods where residents engaged in work, home life, and leisure in close proximity (Neilson 1995).

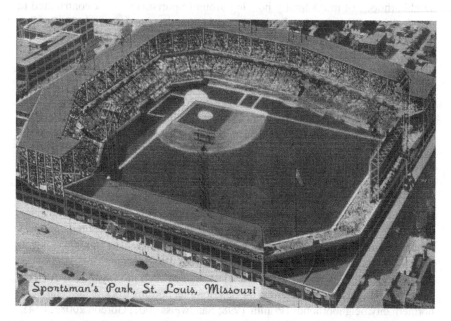

Sportsman's Park, St. Louis, Missouri

Figure 9.2 A Sportsman's Park postcard, circa 1950. The aging stadium, deteriorating surrounding neighborhood, and low automobile accessibility encouraged team
owners to construct a new stadium in the 1960s.

Originally produced by Colourpicture Publishers Inc., Boston, MA and reproduced from the author's
private collection

Early 1900s professional baseball stadiums were not typically attractive,
amenity-laden places. Frequently described as *minimalist* or *utilitarian*, the first
permanent, professional parks were outdoor stadiums constructed solely for baseball played on natural grass (Neilson 1995; Fairfield 2001; Ritzer and Stillman
2001). Easily accessible to the public of rapidly urbanizing cities, stadiums were
park-like settings open to their neighboring city that echoed the striking monumental landscapes and public spaces of the City Beautiful movement (Fairfield 2001).
The classic era stadiums were eventually surrounded by working-class neighborhoods whose fans walked or used public transportation in the form of electric
streetcars, subways, or elevated trains to travel to games (Neilson 1986, 1995).
Their unique architecture and distinctive layout, location in functioning neighborhoods, and urban landmark status helped foster civic pride and shape a unique
sense of place between urban residents and these idiosyncratic stadiums (Neilson
1986; Raitz 1995b; Fairfield 2001).

Busch Stadium II (1966–2005)

As Sportsman's Park aged, St. Louis Cardinals officials considered building a
new stadium. A slowly deteriorating structure with a small seating capacity, inadequate parking and automobile-based transportation infrastructure, and decaying

neighborhoods of multi-family housing around Sportsman's Park contributed to that decision (City of St. Louis Planning and Urban Design Agency 1950; Bradley 2013). Beginning in the 1930s, population changes began buffeting St. Louis. New home construction focused on the western edge of the city, while in older residential areas (including those surrounding Sportsman's Park) urban decay prevailed. These *zones of attrition* clearly exhibited low levels of maintenance (Kersten and Ross 1968; Primm 1998; Bradley 2013; Gordon 2014).

By the 1950s, increasing disinvestment, a declining tax base, and the continued depopulation of the City of St. Louis led to the national perception that the city was beset by pervasive dereliction and social decay (Bradley 2013). Corporations, jobs, shopping, and people were decentralizing to growing western suburbs. A major downtown office building had not been built in St. Louis since 1929, and an expanding area of the city appeared decrepit and blighted (Kersten and Ross 1968; Bradley 2013; Gordon 2014; O'Neil 2014). Like other Midwestern industrial cities, mid-century economic stagnation made the 1950 U.S. Census the population peak for the City of St. Louis (with more than 850,000 people) as over 100,000 residents out-migrated during the 1950s (Kersten and Ross 1968; Primm 1998; Sandweiss 2001; Bradley 2013). As white, middle-class residents left the city, poorer African-American families replaced them in the deteriorating housing stock of the central and northern city neighborhoods (Primm 1998; Sandweiss 2001; Gordon 2008, 2014).

The reimaging of downtown accelerated in the 1950s as city leaders optimistically envisioned a prosperous urban core with interstate connections to the suburbs; an abundance of government, finance, and private-sector jobs; multiple shopping opportunities; the Gateway Arch and riverfront entertainment; as well as sporting events in a new downtown stadium (City Plan Commission 1960). In 1958 Charles Farris, Director of the Land Clearance and Redevelopment Authority for the City of St. Louis, proposed a new stadium as part of a large urban renewal project that would encourage suburban residents to return downtown for entertainment and other leisure activities (Cowan 2005; O'Neil 2013a). Construction on the $24 million, mostly privately financed, stadium began in 1964 after more than a year of demolition of condemned structures. The first game was played in May 1966 (Reidenbaugh 1987; O'Neil 2013a). Baseball seating capacity was initially more than 46,000 seats, although later renovations allowed for between 50,000 and 57,000 spectators for most of the lifespan of the facility (Reidenbaugh 1987).

Busch Stadium II (originally Civic Center Busch Memorial Stadium), four adjacent parking garages, a hotel, and office buildings covering 31 city blocks were an urban renewal project intended to revive the southern sector of the St. Louis urban core (City Plan Commission 1960; Primm 1998; Sandweiss 2001; Gordon 2008; Bradley 2013; O'Neil 2013a). The project followed the tenets of modernist redevelopment by using superblocks, monumental architecture, entertainment facilities, and abundant parking structures to reshape the derelict industrial, waterfront, and abandoned railroad zones (Rubin 2014). Numerous dilapidated buildings, warehouses, a strip club, boarding houses, and the city's small Chinatown were condemned and demolished using eminent domain (Ling 2004; O'Neil 2013a, b). Coupled with the construction of the Gateway Arch (completed in 1965) on riverfront acreage cleared of derelict warehouses, city leaders were optimistic that they had reversed urban decline (Primm 1998; Sandweiss 2001; Bradley 2013; O'Neil 2014).

The stadium was constructed in the late modern era of the mid-1900s when new facilities were criticized as placeless "look-alike concrete doughnuts" that "lost all semblance of unique character" and lacked aesthetic appeal (Raitz 1995a, ix; see also Ritzer and Stillman 2001; Cowan 2005). It was one of 14 baseball super-stadiums built in the 1960s and 1970s that were designed for mass audiences, constructed outside of existing neighborhoods, and were visually closed to the surrounding city. As well, these new stadiums featured artificial turf, expansive seating, and a circular design that maximized the ability to host multiple sporting events (primarily baseball and football games) throughout the year (Neilson 1986, 1995; Fairfield 2001; Ritzer and Stillman 2001; Cowan 2005). Except for 96 distinctive arches that referenced the nearby Gateway Arch and composed the partial roof circling the top of Busch Stadium, the facility was nearly identical to other new stadiums in Atlanta, Cincinnati, Philadelphia, and Pittsburgh (Figure 9.3). The construction of new multipurpose stadiums that were functional but exuded blandness while contributing to the feeling of inauthenticity were the epitome of geographer Edward Relph's critique of increasing placelessness (1976).

Figure 9.3 A Busch Stadium II postcard with aerial perspective, 1967. The back of the postcard states that the image was "a panorama of St. Louis Progress." Although the 96 arches along the roofline were notable, the multipurpose stadium was closed to and separated from the urban core by expanses of parking lots and garages.

Originally produced by the St. Louis Color Postcard Co., St. Louis, MO and reproduced from the author's private collection

Isolated from surrounding neighborhoods and the perceived dangers of the city by four sizable parking garages and multiple surface parking lots, the site of Busch Stadium II emphasized automobile accessibility with close proximity to U.S. Highway 40 (now Interstate 64) that connected the urban core to the expanding western St. Louis suburbs. Still, Busch Stadium II joined other new stadiums as "urban status symbols" for major league cities, even as they were criticized as placeless multipurpose facilities whose circular shapes disconnected fans from the social geography surrounding the stadium (Neilson 1995, 61). Middle-class consumers commuted from their homes in the suburbs to urban cores for games, typically walking quickly from their cars to the safe and controlled space of the stadium without interacting with the surrounding city or its inhabitants (Cowan 2005).

Decade by decade, the population of the City of St. Louis contracted and suburban St. Louis gained a larger share of the region's population and employment opportunities (Primm 1998; Sandweiss 2001; Bradley 2013; Gordon 2014). Demographic shifts accompanied these changes as nearly 60 percent of the white population out-migrated between 1950 and 1970 and segregation of housing, employment, and social services increased (Gordon 2008; Bradley 2013). As Donald Meinig bluntly summarized in his discussion of American regional structure circa 2000, "in St. Louis, the graceful Gateway Arch frames another display of extensive decline in what was long the southwest anchor of the American Core" (2004, 284). It was once again time to reshape the urban landscape of St. Louis.

Busch Stadium III (2006-present)

Although Busch Stadium II attracted baseball and football fans to downtown St. Louis, at the turn of the century city leaders still sought to reverse the long-standing economic decline and rebuild the social fabric of downtown. City population fell to just under 350,000 by 2000, as the suburban counties in the metropolitan area continued their 50-year long trend of attracting the majority of regional population and economic growth (Sandweiss 2001; Laslo *et al*. 2003; Gordon 2014). In St. Louis City racial segregation and poverty intensified (Sandweiss 2001; Laslo *et al*. 2003; Gordon 2008, 2014). Encouraging functional neighborhoods, employment stability, and lively downtown mixed-use districts became a priority. Sport played a key role in this effort.

The current, retro-style $411 million baseball stadium opened in April 2006 on a site partially in the footprint and immediately south of Busch Stadium II. Architects designed the baseball-only facility to feature a natural grass field, 46,000 seats, an exterior red brick facade, as well as dramatic views of downtown and the Gateway Arch (Figure 9.4). Private donors financed 80 percent of Busch Stadium III, although the project received multiple real estate tax abatements, other tax waivers, as well as government assistance with infrastructure costs (Hunn 2010; Click 2014). Developers modeled the project after Baltimore's Oriole Park at Camden Yards and Cleveland's Progressive Field in the Gateway District, retro-designed, publically funded baseball stadiums surrounded by revitalized central

Figure 9.4 Unlike its predecessor, Busch Stadium III features memorable views of downtown St. Louis and the Gateway Arch, maintaining a visual link between baseball games and the adjacent city.

Photo by author, 2014

city regions focused on sport, tourism, and consumption. Baseball-only stadiums with natural grass fields, unique architectural details, a downtown location surrounded by other businesses, a design visually open to the city skyline, and a smaller seating capacity (typically in the mid-40,000s) characterizes postmodern ballparks constructed beginning in the 1990s (Neilson 1995; Ritzer and Stillman 2001; Chapin 2004).

Unique stadiums can quickly become landmarks when they complement the surrounding urban landscape. By integrating within their respective city, the new stadiums enhance place-making and can serve as a centerpiece of urban renewal projects (Sheard 2001). Other cities including Cincinnati, Denver, Detroit, Pittsburgh, San Francisco, and Seattle have embraced architecture reminiscent of the classic period and constructed new central city stadiums (Newsome and Comer 2000; Rosensweig 2005; Click 2014). Increasingly, retro stadiums emphasize a safe leisure experience with luxury boxes, restricted all-inclusive areas, as well as multiple opportunities for consumption in the ballpark including shopping, fine dining in luxury boxes, and play space for children (Fairfield 2001; Ritzer and Stillman 2001; Rosensweig 2005; Gordon 2013). As observers of urban sports history have critiqued, new retro-style stadiums are constructed to evoke nostalgia for the past but function much like suburbs (Fairfield 2001) and shopping malls (Ritzer and Stillman 2001; Friedman 2016).

The agreement that secured public funding for Busch Stadium III mandated that the Cardinals develop Ballpark Village (an entertainment-commercial-residential district situated on a 10-acre, seven city block site immediately north of the stadium). The Ballpark Village proposal was aggressively promoted by city officials as an important catalyst for downtown economic renewal and played a significant role in securing public funding for the stadium (Ponder 2004; Patton 2008). Original proposals by the Cardinals called for a two-phase project, with commercial space scheduled to open in 2007 and the entire $650 million development completed by 2011 (Gose 2013; Click 2014). However, the beginning of the Great Recession in 2007 delayed Ballpark Village construction and the property unbecomingly housed a dirt pit, then a softball field, and finally a large parking lot (Crisp 2012; Click 2014).

The Cardinals and their development partners (Cordish Companies) revised the original Ballpark Village plans and completed a simplified, $100 million, two city block phase of the project devoted to entertainment in March 2014. State and city incentives contributed $17 million to the cost (Logan 2013). Later stages of the Ballpark Village project hint at a New Urbanist design and are projected to include office space, a residential apartment tower, a hotel, and more retail shops with the purpose of creating a pedestrian-friendly, mixed-use neighborhood. A sizable parking lot remains on the ground set aside for future development, although in October 2016 the Cardinals announced a timetable for the $220 million second phase of the project (Bryant 2014a, b; Bryant 2015, 2016).

Ballpark Village features abundant entertainment options with Fox Sports Midwest Live! (a large restaurant serving sports-bar food in front of an enormous 35-foot wide television), a three-story Budweiser Brew House, Cardinals Nation (home to the Cardinals Hall of Fame and Museum and rooftop seats with views into the stadium), and a live event plaza anchoring the space (Hartwig 2014). Ballpark Village quickly proved to be popular destination even on non-game days, with more than four million visitors in 2014 and nearly seven million in 2015 (Bryant 2015).

The (re-)transformation of the southern urban core is part of a larger remaking of the urban identity of St. Louis. Severe population decline and economic loss has led civic leaders to create new consumptive spaces, in this case a new stadium surrounded by baseball-themed entertainment. Numerous home games (81) and a long season (six months) provide many opportunities for fans to frequent nearby downtown shops and restaurants. In addition to constructing a new, instant landmark, the current Busch Stadium and Ballpark Village project seeks to utilize investment to shape a new, post-industrial image of St. Louis as a revitalized city and regional tourist and entertainment destination.

Advocates of sports-based urban development argue that stadiums and their adjacent entertainment districts anchor downtowns and act as catalysts for additional economic investment. Proponents of Ballpark Village argue that the project is already encouraging additional commercial and retail development in downtown St. Louis, including the conversion of old buildings into apartments and condominiums as well as the renovation of the Mercantile Exchange (MX) entertainment district six city blocks north of Ballpark Village along gentrified Washington Avenue (Gose 2013). However, detractors argue that little additional spending has been created, but instead fans have realigned their consumption from elsewhere in the region. Multiple downtown restaurants and bars have closed since 2014 and their owners have argued that loss of clientele to Ballpark Village restricted their ability to remain economically viable (Pistor 2015; Johnson 2016).

Conclusions and implications

Increasingly, politicians and developers in American cities are reallocating additional space to stadiums and sports-related enterprises. Civic leaders of formerly industrial cities are transforming their landscapes to emphasize tourism, entertainment, and sport as they craft new urban identities and promote positive place images. In St. Louis, baseball has a central position in the urban tourist and entertainment economies.

Some observers argue that tourist bubbles and enclaves are becoming increasing homogenized spaces, characterized by standardized entertainment-oriented businesses run by placeless multinational corporations that replace locally owned shops as gentrification removes long-term residents (Judd 1999; Friedman *et al.* 2004; Gotham 2005). Michael Silk and John Amis argue that new sporting venues "have little or no connection to the city" which contributes "to the loss, or indeed palpable weakening, of marks of history, tradition and distinction" (2005, 294). Other studies concur that sports, and by extension new stadiums, have little ability to enhance community bonds in their host cities (Eckstein and Delaney 2002; Smith and Ingham 2003). Observers have also argued that the positive economic impacts of sports infrastructure are overstated and unsustainable while criticizing urban sports-based tourist enclaves for promoting uneven development, fostering growth in a small region while ignoring economic and social challenges in the vast majority of an urban area (Riess 1989; Baade 1996; Campbell 1999; Newsome and Comer 2000; Siegfried and Zimbalist 2000; Fairfield 2001; Austrian and Rosentraub 2002; Miller 2002; Chapin 2004; Hall 2004; Cowan 2005; Rosensweig 2005). Certainly, the construction of Busch Stadium III and Ballpark Village has not reversed the trend of declining population in St. Louis City. Between 2000 and 2010, the city lost nearly 29,000 people (for a total population of just under 320,000 – a drop of more than 60 percent from the 1950 peak) (City of St. Louis 2011).

Alternately, one can argue that the recent proliferation of downtown, retrostyle, single-sport stadiums surrounded by mixed land use is a welcome aesthetic change from the 1960s model of multi-sport facilities surrounded by a homogenous land use. Many fans and observers view new, retro-style baseball stadiums as authentic places that link the present to baseball's heritage (Friedman *et al.* 2004; Ramshaw and Gammon 2005; Gordon 2013). While hard to quantify, the perceptions of authenticity and inauthenticity – or place and placelessness – is a significant component of a person's willingness to attach emotional feelings to their experience in a sports landscape. Sports studies scholar Kiernan Gordon argues that distinctive ballparks that integrate with their surrounding urban landscape "serve as place anchors," heightening attachment to place due to the memories created, emotions invested, and personal bonds formed while attending sporting events (2013, 232; see also Bale 1996). As urban planners and stadium developers continue to engage in the process of sports-based place-making, the belief that sport fosters community pride, creates readily-identifiable urban landmarks, and makes central cities places to work and live once again is difficult to ignore (Campbell 1999; Hall 2004; Ponder 2004; Ramshaw and Gammon 2005; Smith 2005; Schimmel 2006; Cantor and Rosentraub 2012). Contemporary and historic sports stadiums are significant repositories of our collective memory, facilitating more intimate connections with place.

Although the economic impacts of professional sports infrastructure are debatable, more American cities are turning to stadium development and sports entertainment to reinvent their downtowns and reshape their urban images. Meanwhile, many fans view baseball stadiums as symbolic sites of meaning, facilitating emotional connections with their community and linking them to a nostalgic past. While other urban economic development options remain limited in a globalizing world and sports loyalties continue to be a repository of shared community meaning, no doubt the reshaping of American urban landscapes through sport will continue.

References

Archer, K. 1997. The Limits to the Imagineered City: Sociospatial Polarization in Orlando. *Economic Geography* 73: 322–336.

Austrian, Z. and M. S. Rosentraub. 2002. Cities, Sports, and Economic Change: A Retrospective Assessment. *Journal of Urban Affairs* 24: 549–563.

Baade, R. A. 1996. Professional Sports as Catalysts for Metropolitan Economic Development. *Journal of Urban Affairs* 18: 1–17.

Bale, J. 1988. The Place of "Place" in Cultural Studies of Sports. *Progress in Human Geography* 12: 507–524.

Bale, J. 1996. Space, Place and Body Culture: Yi-Fu Tuan and a Geography of Sport. *Geografiska Annaler: Beries B, Human Geography* 78: 163–171.

Bale, J. 2003. *Sports Geography*, 2nd ed. New York: Routledge.

Bradley, B. H. 2013. *Thematic Survey of Modern Movement Non-Residential Architecture, 1945–1975, in St. Louis City, Historic Context Statement: St. Louis: The Gateway Years, 1940–1975*. St. Louis: City of Saint Louis Cultural Resources Office.

Bryant, T. 2014a. DeWitt Says Ballpark Village Suitable for Office and Residential Towers. *St. Louis Post-Dispatch*. www.stltoday.com/business/columns/building-blocks/dewitt-says-ballpark-village-suitable-for-office-and-residential-towers/article_74bbc053-da0c-5050-8e9b-79f987d35740.html. Last accessed 1 September 2015.

Bryant, T. 2014b. Opening at Last . . . The $100 Million First Phase of the Long-Delayed Project Is Sure to Make a Splash When It Opens, but St. Louisans Are Already Looking Ahead to What's Next. *St. Louis Post-Dispatch*, pp. A1+A8, 23 March.

Bryant, T. 2015. Second Season. *St. Louis Post-Dispatch*, pp. B1+B4, 27 March.

Bryant, T. 2016. Cardinals Reveal Expansion Plans. *St. Louis Post-Dispatch*, pp. A1+A6, 26 October.

Campbell, Jr., H. S. 1999. Professional Sports and Urban Development: A Brief Review of Issues and Studies. *The Review of Regional Studies* 29: 272–292.

Cantor, M. B. and M. S. Rosentraub. 2012. A Ballpark and Neighborhood Change: Economic Integration: A Recession, and the Altered Demography of San Diego's Ballpark District after Eight Years. *City, Culture, and Society* 3: 219–226.

Chapin, T. S. 2004. Sports Facilities as Urban Redevelopment Catalysts: Baltimore's Camden Yards and Cleveland's Gateway. *Journal of the American Planning Association* 70: 193–209.

City Plan Commission. 1960. *A Plan for Downtown St. Louis*. St. Louis: City Plan Commission, City of Saint Louis. City of St. Louis website. www.stlouis-mo.gov/government/departments/planning/documents/a-plan-for-downtown-st-louis-1960.cfm. Last accessed 10 January 2016.

City of St. Louis Planning and Urban Design Agency. 1950. Historical Map of St. Louis 1950 Zoning. City of St. Louis website. www.stlouis-mo.gov/government/departments/planning/documents/historical-map-of-stl-1950-zone.cfm. Last accessed 10 January 2016.

City of St. Louis Planning and Urban Design Agency. 2011. St. Louis City 2010 to 2000 Comparison Census Report. City of St. Louis website. http://dynamic.stlouis-mo.gov/census/cen_city_comp.cfm. Last accessed 9 February 2016.

The Civic League of St. Louis. 1907. *A City Plan for Saint Louis*. City of St. Louis website. www.stlouis-mo.gov/government/departments/planning/planning/other-plans/1907-city-plan-for-saint-louis.cfm. Last accessed 16 February 2016.

Click, E. 2014. One Development Project, Two Economic Tales: The St. Louis Cardinals' Busch Stadium and Ballpark Village. *Missouri Policy Journal* 2: 21–34.

Cowan, A. 2005. A Whole New Ball Game: Sports Stadiums and Urban Renewal in Cincinnati, Pittsburgh, and St. Louis, 1950–1970. *Ohio Valley History* 5: 63–86.

Crisp, E. 2012. Ballpark Village to Open in 2014 after Missouri Board Oks Incentives. *St. Louis Post-Dispatch*. www.stltoday.com/news/local/govt-and-politics/ballpark-village-to-open-in-after-missouri-board-oks-incentives/article_8db7559c-e081–5bb2–8192-a98a5c722425.html. Last accessed 24 July 2015.

Doss, E. 2010. *Memorial Mania: Public Feeling in America*. Chicago: University of Chicago Press.

Duquette, G. H. and D. S. Mason. 2008. Urban Regimes and Sport in North American Cities: Seeking Status through Franchises, Events and Facilities. *International Journal of Sport Management and Marketing* 3: 221–241.

Eckstein, R. and K. Delaney. 2002. New Sports Stadiums, Community Self-Esteem, and Community Collective Conscience. *Journal of Sport & Social Issues* 26: 235–247.

Fairfield, J. D. 2001. The Park in the City: Baseball Landscapes Civically Considered. *Material History Review* 54: 21–39.

Friedman, M. T. 2010. "The Transparency of Democracy": The Production of Washington's Nationals Park as a Late Capitalist Space. *Sociology of Sport Journal* 27: 327–350.

Friedman, M. T. 2016. Supercharging the Mallpark: Battery Atlanta and the Future of Baseball Stadium Development. Presentation at the *Annual Meeting of the American Association of Geographers*, San Francisco, California, 1 April.

Friedman, M. T., D. L. Andrews, and M. L. Silk. 2004. Sport and the Façade of Redevelopment in the Postindustrial City. *Sociology of Sport Journal* 21: 119–139.

Gaffney, C. T. 2008. *Temples of the Earthbound Gods: Stadiums in the Cultural Landscapes of Rio de Janeiro and Buenos Aires*. Austin: University of Texas Press.

Gammon, S. 2004. Secular Pilgrimage and Sport Tourism. In *Sport Tourism: Interrelationships, Impacts, and Issues*, edited by B. W. Ritchie and D. Adair, 30–45. Buffalo, NY: Channel View.

Gordon, C. 2008. *Mapping Decline: St. Louis and the Fate of the American City*. Philadelphia: University of Pennsylvania Press.

Gordon, C. 2014. St. Louis Blues: The Urban Crisis in the Gateway City. *Saint Louis University Public Law Review* 33: 81–92.

Gordon, K. 2013. Emotion and Memory in Nostalgia Sport Tourism: Examining the Attraction to Postmodern Ballparks through an Interdisciplinary Lens. *Journal of Sport & Tourism* 18: 217–239.

Gose, J. 2013. St. Louis Development Makes a Play for a Home Team Advantage. *The New York Times*, p. B7, 15 May.

Gotham, K. F. 2005. Tourism Gentrification: The Case of New Orleans' Vieux Carre (French Quarter). *Urban Studies* 42: 1099–1121.

Gripshover, M. 2008. Lake View, Baseball, and Wrigleyville: The History of a Chicago Neighborhood. In *Northsiders: Essays on the History and Culture of the Chicago Cubs*, edited by G. C. Wood and A. Hazucha, 11–26. Jefferson, NC: McFarland & Company.

Gumprecht, B. 2008. *The American College Town*. Amherst: University of Massachusetts Press.

Hall, C. M. 2004. Sport Tourism and Urban Regeneration. In *Sport Tourism: Interrelationships, Impacts, and Issues*, edited by B. W. Ritchie and D. Adair, 192–205. Buffalo: Channel View.

Hartwig, G. 2014. Heading to Ballpark Village? Here's Your Guide. *St. Louis Post-Dispatch*. www.stltoday.com/entertainment/music/heading-to-ballpark-village-heres-your-guide/article_62eb7d21-a46b-5ed3-9061-a82c8d6bfe59.html. Last accessed 25 July 2015.

Hoelscher, S. 2006. Heritage. In *A Companion to Museum Studies*, edited by S. MacDonald, 198–218. Oxford: Blackwell Publishers.

Hunn, D. 2010. Do Cards Owe City in Ballpark Deal? *St. Louis Post-Dispatch*. www.stltoday.com/sports/baseball/professional/do-cards-owe-city-in-ballpark-deal/article_dc308b52-aaa4-5f52-aa30-9f05e6040265.html. Last accessed 30 August 2015.

Jackson, J. B. 1984. *Discovering the Vernacular Landscape*. New Haven: Yale University Press.

John, G. 2002. Stadia and Tourism. In *Sport Tourism: Principles and Practice*, edited by S. Gammon and J. Kurtzman, 53–61. Eastbourne: Leisure Studies Association.

Johnson, K. C. 2016. Co-Owner Pins Harry's Closure on Ballpark Village. *St. Louis Post-Dispatch*, p. A2, 19 January.

Judd, D. 1999. Constructing the Tourist Bubble. In *The Tourist City*, edited by D. R. Judd and S. S. Fainstein, 35–53. New Haven: Yale University Press.

Kersten, E. W. and D. R. Ross. 1968. Clayton: A New Metropolitan Focus in the St. Louis Area. *Annals of the Association of American Geographers* 58: 637–649.

Laslo, D., C. Louishomme, D. Phares, and D. R. Judd. 2003. Building the Infrastructure of Urban Tourism: The Case of St. Louis. In *The Infrastructure of Play: Building the Tourist City*, edited by D. R. Judd, 77–103. London: Routledge.

Lewis, P. 1983. Learning from Looking: Geographic and Other Writing about the American Cultural Landscape. *American Quarterly* 35: 242–261.

Ling, H. 2004. *Chinese St. Louis: From Enclave to Cultural Community*. Philadelphia: Temple University Press.

Logan, T. 2013. Groundbreaking Set for Ballpark Village. *St. Louis Post-Dispatch*, p. A2, 31 January.

Lowenthal, D. 1998. *The Heritage Crusade and the Soils of History*. Cambridge: Cambridge University Press.

Meinig, D. W. 2004. *The Shaping of America: A Geographical Perspective on 500 Years of History, Volume 4: Global America, 1915–2000*. New Haven: Yale University Press.

Miller, P. A. 2002. The Economic Impact of Sports Stadium Construction: The Case of the Construction Industry in St. Louis, MO. *Journal of Urban Affairs* 24: 159–173.

Neilson, B. J. 1986. Dialogue with the City: The Evolution of Baseball Parks. *Landscape* 29: 39–47.

Neilson, B. J. 1995. Baseball. In *The Theater of Sport*, edited by K. B. Raitz, 30–69. Baltimore: The Johns Hopkins University Press.

Newsome, T. H. and J. C. Comer. 2000. Changing Intra-Urban Location Patterns of Major League Sports Facilities. *Professional Geographer* 52: 105–120.

Novak, M. 1976. *The Joy of Sports*. New York: Basic Books, Inc.

O'Neil, T. 2013a. New Home for Cards Revitalized Downtown. *St. Louis Post-Dispatch*, p. B3, 12 May.

O'Neil, T. 2013b. Fading Chinatown Falls to Redevelopment. *St. Louis Post-Dispatch*, p. B3, 4 August.

O'Neil, T. 2014. Downtown Evolves. *St. Louis Post-Dispatch*, p. B3, 28 December.

Patton, E. 2008. *Covering the Bases: Variations in the Arguments to Justify Publicly-Funded Baseball Stadiums*. M.A. thesis, Geography, University of Missouri, Columbia, MO.

Pistor, N. J. C. 2015. Billion-Dollar Effort Could Turn Urban Tide. *St. Louis Post-Dispatch*, pp. A1+A5, 7 October.

Ponder, B. 2004. Ballpark Village or Urban Wasteland: Rhetorical Invention, Civic Ego, and the St. Louis Cardinal's Controversial New Home. *NINE: A Journal of Baseball History and Culture* 13: 74–80.

Primm, J. N. 1998. *Lion of the Valley: St. Louis, Missouri, 1764–1980*. St. Louis: Missouri Historical Society Press.

Raitz, K. B. 1995a. Preface. In *The Theater of Sport*, edited by K. B. Raitz, vii–xiv. Baltimore: The Johns Hopkins University Press.

Raitz, K. B. 1995b. The Theater of Sport: A Landscape Perspective. In *The Theater of Sport*, edited by K. B. Raitz, 1–29. Baltimore: The Johns Hopkins University Press.

Ramshaw, G. and S. Gammon. 2005. More Than Just Nostalgia? Exploring the Heritage/Sport Tourism Nexus. *Journal of Sport Tourism* 10: 229–241.

Reidenbaugh, L. 1987. *The Sporting News Take Me Out to the Ball Park*, 2nd ed. St. Louis: The Sporting News Publishing Company.

Relph, E. 1976. *Place and Placelessness*. London: Pion Limited.

Riess, S. A. 1989. *City Games: The Evolution of American Urban Society and the Rise of Sports*. Chicago: University of Illinois Press.

Ritzer, G. and T. Stillman. 2001. The Postmodern Ballpark as a Leisure Setting: Enchantment and Simulated De-McDonaldization. *Leisure Sciences* 23: 99–113.

Rosensweig, D. 2005. *Retro Ball Parks: Instant History, Baseball, and the New American City*. Knoxville: University of Tennessee Press.

Rosentraub, M. 2014. *Reversing Urban Decline: Why and How Sports, Entertainment, and Culture Turn Cities in Major League Winners*, 2nd ed. Boca Raton: CRC Press.

Rubin, J. 2014. Planning and American Urbanization since 1950. In *North American Odyssey: Historical Geographies for the Twenty-First Century*, edited by C. E. Colten and G. L. Buckley, 395–411. Lanham, MD: Rowman & Littlefield Publishers.

Sandweiss, E. 2001. *St. Louis: The Evolution of an American Urban Landscape*. Philadelphia: Temple University Press.

Schimmel, K. S. 2006. Deep Play: Sports Mega-Events and Urban Social Conditions in the USA. *Sociological Review* 54: 160–174.

Sheard, R. 2001. *Sports Architecture*. New York: Spon Press.

Short, J. R. 1999. Urban Imagineers: Boosterism and the Representation of Cities. In *The Urban Growth Machine: Critical Perspectives Two Decades Later*, edited by A. E. G. Jonas and D. Wilson, 37–54. Albany: State University of New York Press.

Siegfried, J. and A. Zimbalist. 2000. The Economics of Sports Facilities and Their Communities. *The Journal of Economic Perspectives* 14: 95–114.

Silk, M. and J. Amis. 2005. Sport Tourism, Cityscapes and Cultural Politics. *Sport and Society* 8: 280–301.

Smith, A. 2005. Reimaging the City: The Value of Sport Initiatives. *Annals of Tourism Research* 32: 217–236.

Smith, J. M. and A. G. Ingham. 2003. On the Waterfront: Retrospectives on the Relationship between Sport and Communities. *Sociology of Sport Journal* 20: 252–274.

Stone, G. 1981. Sport as Community Representation. In *Handbook of Social Science of Sport*, edited by G. R. F. Lüschen and G. H. Sage, 214–245. Champaign, IL: Stipes Publishing.

Trumpbour, R. 2007. *The New Cathedrals: Politics and Media in the History of Stadium Construction*. Syracuse: Syracuse University Press.

Turner, R. S. 2002. The Politics of Design and Development in the Postmodern Downtown. *Journal of Urban Affairs* 24: 533–548.

Turner, R. S. and M. S. Rosentraub. 2002. Tourism, Sports and the Centrality of Cities. *Journal of Urban Affairs* 24: 487–492.

Zelinsky, W. 2011. *Not Yet a Placeless Land: Tracking an Evolving American Geography*. Amherst: University of Massachusetts Press.

10 *Avant-garde*, wannabe Cowboys

Place attachment among Bohemians, Beatniks, and Hippies in Virginia City, Nevada

Engrid Barnett

Located high in the Virginia Range running parallel to the eastern slope of the Sierra Nevada, with a forbidding high desert climate and challenging mountainous terrain, the remnants of nineteenth-century Virginia City, Nevada, presented the ultimate soundstage for the re-creation of America's frontier heritage to suit the needs of a new generation of twentieth-century Americans who had no vast tracts of free land to tame. Rising 6,200 feet above sea level, Virginia City (also referred to as the *Comstock* after one of its original prospectors, Henry T. "Pancake" Comstock) was originally thought to be sandwiched between the glittering veins of Eldorado below and unobstructed glimpses of heaven above, where the viewer's eye could reach out and capture at least 50 miles of landscape. A lonely space highlighted by limitless silence, skies of flawless periwinkle, and boundless personal freedom, the twentieth-century Comstock was a place that imbued visitors with a sense of America's greatness through the legends that they peddled of Wild West forefathers and mothers (Hausladen 2003). Clustered like a broken strand of pearls, the remaining historic buildings offered quaint testimonies to a bygone age and a reflection of the area's most stubborn individuals, those who managed to construct a thriving urban center despite tremendous natural obstacles.

The real treasure of the Comstock lay not only beneath the feet of the community but in the well-trodden myths of the Wild West with which this fading community would be associated. By the end of the nineteenth century, the fabled silver mines of the Comstock had nearly played out, and the city-turned-town-turned-quasi-ghost-town was near death. Enterprising business owners in the area, however, resuscitated the once glorious mineral district by tapping into the well-spring of tourism. This mighty task would have proven impossible were it not for a handful of eclectic Bohemians who moved West in search of the last glimmers of the Turnerian past. Together, the city's business owners and relocated Bohemians managed, in a space of two decades, to render Virginia City emblematic of the American West. This process, however, alienated many in the latter group as the city's sense of place ultimately suffered. Nonetheless, this feat led to other great counterculture migrations, namely by Beatniks and Hippies, whose presence and influences remain readily tangible to this day.

This chapter highlights the unique character of a place where two waves of counterculture residents have developed deep emotional ties to the historic mining city.

The character of the place validates the lives of these latter-day emigrants seeking escape from the constraints of mainstream culture, morality, and politics. Information for this chapter was derived from historic newspaper articles, journals, photographs, and firsthand accounts of key actors as well as secondary literature including scholarly books, articles, and biographies. Additionally, I conducted an in-depth examination of the cultural landscape aided by my proximity to the area and the many eyewitnesses and participants still able to share their insights. As the research took on the character of an extended oral history project with over 50 formal and informal interviews conducted, I sought and received IRB approval. This required developing research contacts, preparing interview questions, conducting in-person and phone interviews, and processing the resulting digital voice files into transcripts. The interviewing process comprised, by far, the largest chunk of research that I undertook and delivered the greatest dividends.

Virginia City's rich history of tourism

The American West has always been inextricably tied to tourism (Rothman 1998; Barber 2008). The events that Easterners, Europeans, and various individuals from around the globe read about in newspapers, guides, and books from the various regions of the Far West were simply too enticing *not* to draw crowds, and Virginia City benefited from this attraction, nearly from its point of inception (Hausladen 2003). As early as 1860–61, infamous travel writer, artist, and author, J. Ross Browne, spun a sojourn to the Comstock into rough-and-ready, hyperbole-laden frontier articles featured in *Harper's Magazine* (see: Browne 1860–61). Waves of tourists followed in Browne's footsteps: "[They] came to see the mighty Comstock, home of the big bonanza . . . Virginia City was a must-see stop on any excursion through the region" (James 1998, 238). Browne's articles offended countless Comstock residents while launching a new strain of Western literary mythmaking. Authors such as Bret Harte, Mark Twain, and Dan DeQuille continued in Browne's wake, crafting exaggerated tales of the colorful existence enjoyed out West (DeQuille 1974; Twain 2008). Tourists trekked to the area to ride the Virginia and Truckee (V&T) Railroad, which offered long vistas and a hair-raising journey across the Crown Point Trestle, "one of the engineering wonders of the West" (Beebe and Clegg 1949, 54–55; James 1998, 82). A structure spanning 500 feet across the Crown Point Ravine and elevated 85 feet off the ground, its image became so iconic to Nevada history that Comstockers erroneously claimed that the trestle featured on Nevada's state seal was their impressive Gold Hill jewel.

On October 27th, 1879, former U.S. President, Ulysses S. Grant, and his family traveled to Virginia City where they passed three days, showered with every gift, experience, and sight that the area could offer. The city celebrated their every move with the pomp and pageantry usually reserved for the Fourth of July. President Hayes, his wife, and an entourage of cabinet members soon followed Grant's example as did countless nineteenth-century celebrities and notables.

Besides flaunting its great industrial might, the nineteenth-century community knew how to exploit the dangerous, exotic, racist, and terrifying to capture the imagination, and part a man from his money (Peck 1993). Salacious entertainments such as Native American vaudeville, bare-knuckle boxing, and unmatched animal fights (e.g., a bear versus dogs) were interspersed with sensational news stories meant to sell papers and books (Cafferata 2008; James 1998; Watson 1964). The Wild West went on to great infamy, the myth of regeneration through violence firmly rooted in the American imagination.

As Hal Rothman (1998) points out, tourism brings with it certain tensions, particularly between the development of a residential sense of place and the needs of the tourist industry. In the case of Virginia City, however, any type of residential sense of place was more or less in tatters by the early twentieth century. Many members of the community felt they had nothing to lose, and so Comstockers salvaged lumber from abandoned buildings to warm their own homes through cold winter months. The town lamentably and significantly dwindled.

> The original builders of Virginia City had placed many houses side by side, as they are yet today in the older parts of San Francisco. The Depression-era strategy of house demolition gave homeowners more space and was consistent with the changing ideas of domestic land use in the twentieth century.
> (James 1998, 252)

Depression-era cannibalization of the city was so successful that the few houses that remain side-by-side (as they were originally built) appear rare and remarkable today. To explain the peculiarity of the remaining structures, mid-twentieth century residents invented the tale of the "spite houses." According to legend, two neighbors in a dispute over property lines built their houses so close together (to each lay their claim) that they appear to be kissing in spite of themselves. In less than a century, an oral tradition sprang up to replace actual memories of the original city's construction.

Despite the cost of the "devil's bargain" that Virginia City residents made, they had precious few options. The town faced extinction. "Regions, communities, and locales welcome tourism as an economic boom, only to find that it irrevocably changes them in unanticipated and uncontrollable ways" (Rothman 1998, 17). For Virginia City, much of this change came in the form of rewritten city histories more closely adhering to Hollywood's portrayal of the Wild West. Starting as early as the 1930s and 1940s with the arrival of a handful of well-intentioned, albeit sentimental, Bohemians and intellectuals, civic folklore was preserved, expanded, and canonized. They actively sought remnants of the Wild West refashioning Virginia City to meet preconceived, romanticized expectations. Some, like Lucius Beebe and Charles Clegg, emphasized authenticity. Some, including Walter Van Tilburg Clark, wrestled with the significance of the frontier in a post-Turnerian landscape. Others, like Duncan Emrich, spun a web of "fakelore" buttressing the town's economic livelihood to this day (Dorson 1977).

Virginia City's landscape made its citizens desperately believe in Turner's egalitarian Wild West of rugged individuals. By no means did Turner singlehandedly invent the frontier concept, though. Rather, he reoriented the history of the United

States based on what conventional wisdom wanted and needed to believe – that "the pulse of the nation lay in its frontier and was defined by evolving settlement patterns as gleaned from data in the 1890 United States Census" (Turner 1891, 1961). If such were true, reasoned Comstockers, then the heart from which that pulse beat must be Virginia City.

Quaint, rundown buildings and the tourist's version of fool's gold (e.g., costume-clad desperados, sheriffs, and soiled doves) fueled the avarice-driven descent into fakelore. The contradictions between fantasy and reality could be easily mini-mized through the cultivation of a particular appearance aided by decades of decay, dilapidation, and, in some cases, construction of Hollywood-inspired travesties. Business owners had but to add a tinge of suggestion, and tourists' imaginations spun wildly out of control. Where else but C Street would gunslingers race down the wooden boardwalk firing wildly and haphazardly, before ducking into saloons? Where else but D Street would famed prostitutes the likes of Julia Bulette assert feminine mastery over a community of rough-and-tumble males through sensual-ity, sex, and sophistication? Where else but B Street would lonely, old prospectors with heavily-laden mules commiserate over what they'd lost and all they yet hoped to gain from fickle Mount Davidson? Ultimately, myth sells.

By the mid-twentieth century, gone were many of the Comstock's chief indus-trial marvels and more permanent features of community (James 1998). The deep-mine pumps, stamp mills, shafts, and adits no longer rang, pounded, and clattered with "progress" while incessantly belching heavenward smoke, toxins, pulverized ore, and chaos (Brechin 2001; Lord 1959; Moehring 1997). The V&T railroad no longer clanged, whistled, and steamed into Virginia City's depot on D Street. St. Mary's Hospital on R Street, a state-of-the-art facility constructed in 1875 to treat victims of industrial accidents, stood in neglect isolated from the town's downtown corridor. The six-story International Hotel, an elegantly turned out brick building with access to C Street (main street after the 1875 fire) and the entrance to Piper's Opera House on B Street (after 1885) had long since burned to the ground. Ironically, what made Virginia City a working industrial town of the nineteenth century (e.g., noise, pollution, industrial equipment, sewage, vast armies of laborers) held little interest for twentieth-century tourists. So, adjust-ments to municipal history liberally occurred. Comstock entrepreneurs served up the most salacious and narrow-minded iconographies of the mediatized Wild West and met with economic success.

Bohemian arrivals codify Virginia City's myths

The 1930 U.S. Census reports 667 people living in Virginia City and its Storey County environs. The Comstock was suffering from severe population decline when a small tribe of Bohemians and intellectuals arrived.

> [B]eginning as early as the 1930s a strange transformation occurred on the Comstock. What had been a dormant, even dying, mining district slowly blos-somed into a magnet for artists, literati, and others who wished to experience

something of the fast-disappearing Wild West. They were attracted to Virginia City because they believed that the place had not changed. Of course it had.

(James 1998, 258–259)

Cheap rent, scenic vistas, proximity to Reno and San Francisco, and distance from the watchful eyes of law enforcement proved irresistible siren calls for sub-culture adherents who effortlessly attracted kindred spirits (Ullman 1954). "The presence of a significant Bohemian concentration signals a regional environment or milieu that reflects an underlying openness to innovation and creativity. This milieu is both open to and attractive to other talented and creative individuals" (Florida 2002, 56). During Prohibition, the absence of law enforcement combined with brazen local bootleggers and inventive speakeasies kept Virginia City any-thing but dry. This tolerant landscape proved particularly alluring for Bohemi-ans who came as tourists, and took up residence in droves. Lured by Virginia City's makeover into the Tombstone/Deadwood of Nevada, Bohemians altered their dress, behaviors, and lifestyles to harmonize with the historical and cultural landscapes of these aggrandizing place images thereby encouraging an intense connection and attachment to place. Ultimately, the land *validated* key aspects of their patriotism, identity, and understanding of history.

Post-World War II America – the "Age of Anxiety" – desperately needed its fictions. The Wild West with its larger-than-life heroes and villains sold while reinforcing a sense of virile manliness that thrived on the battlefields of WWII but found few outlets in 1950s suburbia. In the pioneer woman, 1950s housewives re-connected with the independence, adventure, and hardiness that they briefly enjoyed from 1941–45 when they shouldered the burden of America's hefty war effort. The children of the bewildering Atomic Age found refuge in Western "fai-rytales" with distinct values and undeniable heroes. The Turnerian narrative also allowed for black-and-white contrasts to be made between "virtuous" American capitalism and "depraved" Communist depotism. The nation's desire to highlight individualism and single-minded exceptionalism over and above actual demon-strations from the documentary record of what communities were able to achieve in synch, is clearly discernible.

One of the more colorful members of the Comstock community would be the cosmopolitan scholar, Duncan Emrich, whose prestigious degrees from Brown, Columbia, and the University of Madrid helped him attain the position of founding director of the Folklore Section of the Library of Congress before his fortieth birth-day. "Like so many of his Bohemian counterparts, he thirsted to promote the Com-stock's unsavory fakelore" (Dorson 1977, 4). Through dubious interviews at the Delta Saloon, Emrich crafted Julia Bulette (a small-time, crib prostitute murdered in 1867) into the Comstock's most celebrated madam, and he also laid the ground-work for many other local legends. All fell within the context of an especially wild frontier community teetering on the border between civilization and savagery.

Like Emrich, historian and novelist Bernard DeVoto's enduring devotion to the American West brought him to the Comstock, and his popular histories attempted to capture the last flickers of a bygone age: *The Year of Decision: 1846* (1943),

Across the Wide Missouri (1947), and *The Course of Empire* (1952). In 1948, he received the Pulitzer Prize for History for *Across the Wide Missouri*, and he continued to garner acclaim after publishing an abridged version of Lewis and Clark's famed travel journals (1953). But while DeVoto committed himself to preserving a dying and glorified past, others wished the impossible. Beebe and Clegg, East Coast dandies of impeccable style and wit, rode into town with the expressed intention of resuscitating the community.

Avid train aficionados, they relocated from New York's café society in 1950 to revitalize the *Territorial Enterprise* and preserve what they could of a dying age, that of steam locomotion in the West. In the process, Beebe evolved into the most prolific writer on railroading in the United States. With Charles Clegg (avid photographer and Beebe's lifelong partner) they invented the railroad book as a genre. Clegg's glossy images complemented Beebe's engaging prose and ignited many Americans' passion for trains. The medium is as popular today as when they launched it from the Comstock. Their talents went to good use in shining the international spotlight on the V&T Railroad.

Pivotal to bringing Virginia City back into the national spotlight, Beebe and Clegg did so in spectacular style. Beebe reintroduced fine dining and luxurious style by commissioning the renovation of old V&T dining cars into spaces of opulent decor and gourmet excess. One was even converted into a palatial Turkish bath. "[T]the new 'frontier' of America [had become] a frontier of comfort, in contrast with the traditional frontier of hardship" (Ullman 1954, 120). Within two years of purchasing the *Territorial Enterprise* in 1952, the newspaper enjoyed renewed status as a very popular and highly circulated national periodical. The Comstock's rich, nineteenth-century literary tradition (e.g., Dan DeQuille, Mark Twain, Alf Doten) was reborn through their efforts at the *Territorial Enterprise*. Characterized as, "pro-prostitution, pro-alcohol, pro-private-railroad-cars-for-the-few and fearlessly anti-poor folks, anti-progress, anti-religion, anti-union, anti-diet, anti-vivisection and anti-prepared breakfast food," Beebe and Clegg's revamped *Enterprise* suited the long-standing libertarian views of area residents who adhered to one rule: "live and let live" (Gertz 2010, n.p.). Beebe and Clegg brought sumptuousness and attitude to the area while enriching its literature and sophisticating its image.

Nevada's most distinguished literary figure of the twentieth century, however, remains Walter Van Tilburg Clark, author of *The Ox-Bow Incident* (1940), *The City of Trembling Leaves* (1945), and *Track of the Cat* (1949.) He moved to Virginia City in 1949 to teach at the local high school where he worked from 1950–51 and also the University of Nevada, Reno, where he remained until 1953. In 1962, he purchased property in Comstock and took an eighteen-month contract with the university to write a biography of one of Virginia City's great notables, Alf Doten. Obsessed with the compilation and editing of Doten's journals, this project consumed Clark for the remainder of his life and was published posthumously in 1973. He is buried next to his wife in the Comstock's famed Silver Terrace Cemetery.

Numerous other outcast elites and soon-to-be divorcés colored the Comstock's socio-cultural landscape, too, while claiming it as a temporary but necessary home during the waiting period for their divorces.

[Beebe and Clegg] met kindred spirits on their strolls around Virginia City. There were the remittance kids, children of the rich who were paid to stay away. A Delaware du Pont was tending bar at the Sky Deck Saloon. An Eastern socialite was running a hotel. A chef from Maxim's was cooking at the Bonanza Inn. Numerous members of Café Society were on hand, waiting out their six-week residencies at divorce ranches near Reno. . . . Soon pals and colleagues from back East were migrating to Virginia City, and townsfolk took it in good stride while hobnobbing with Cole Porter and other celebrities.

(Daley-Taylor 1992, 22)

These groups of eclectic Bohemians contributed to a sophisticated, creative, dynamic community isolated in the mountains above Reno.

The Comstock became an end in itself, a destination for the jaundiced or bored. In the city's relative isolation and timelessness, they indulged their deepest need for escapism, seeking exile from Cold War current events, an oppressive body politic characterized by fear and conformity, and what was fast morphing into a "plastic" American culture. Artists, authors, and intellectuals found respite from a world gone nearly mad since America's rise to world dominance at the end of World War II. "Many came to the Comstock to sink into the luxurious morass, a thief of time and ambition that served as refuge for internationally prominent ne'er-do-wells fleeing a high-pressure world" (James 1998, 260).

The Comstock, with its reviving Bohemian presence, was clearly poised to surf the nationwide swell in tourism brought about with increased post-war affluence. Rent was extremely cheap, the scenic vistas expansive, and the folklore/place myth decidedly entertaining. Conveniently located within an hour's drive of the divorce capital of the world (Reno), Bohemians and well-to-dos rubbed elbows at local saloons, and new creative communities emerged (Barber 2008). Place attachment was inevitable, aided by the Comstock's capacity to inspire its residents to embody the city's mythologized history.

Tourism and place attachment

Place attachment "refers to the emotional and psychological bonds formed between an individual and a particular place," and it is a theme studied by environmental psychologists and tourism studies researchers, drawing on the works of humanistic geographers to understand the human-land relationship (Relph 1976; Tuan 1974; Brown and Raymond 2007). While psychologists tend to focus on the emotional affectations of place attachment, tourism studies scholars emphasize tangible attributes and activities associated with destinations. Tourism theory models, in particular, attempt to quantify and provide clarity about why tourists are attracted to certain destinations, make repeat visits, and even relocate (Scannell and Gifford 2010). In tourism studies, two theories predominate: the *tourism involvement approach* and the *destination-attribute approach*. The tourism involvement model is particularly well suited to understanding the lure of the Comstock for members of the subculture who came as visitors and remained as residents.

According to the tourist involvement model, individuals develop strong bonds with a place through the evolutionary process of initial attraction, self-expression, and centrality of the place to one's lifestyle (Kyle *et al.* 2004; DeLyser 2005; Hwang *et al.* 2005; Hou *et al.* 2005; Gross *et al.* 2008). These three steps in the process are clearly evident in the case of Virginia City. Its proximity to San Francisco and Reno, legendary folklore, Wild West place imagery and treatment in Hollywood films and television shows attracted Bohemians, Beatniks and Hippies who altered their dress, behaviors, and lifestyles in a form of self-expression inspired by the historical mining district. Moreover, they stayed because they found *validation* of their identities, alternative moralities, and divergent socio-cultural understandings of the world. This is not surprising in the context of the tourism involvement model where "all sorts of dynamic exchanges and contact points in the sensory, affective, social and intellectual dimensions work together to create desirable and memorable experience, out of which the brand/consumer relationships are fostered and strengthened" (Tsai 2012, 139).

Having found a place that met their perceptions and expectations of a frontier community, it is clear from their writing, music, visual art, dress, and other cultural expressions that the new residents developed a strong, even life-altering, place attachment. Clearly, Virginia City attracted colorful characters, boasted dramatic place myths to exploit, and suggested just enough of a naughty reputation for visitors to indulge in taboo-breaking of various stripes. They enjoyed much media coverage as well because, under the stewardship of Beebe and Clegg, the *Territorial Enterprise* was enjoying its highest circulation. Duncan Emrich, Bernard DeVoto, Walter Van Tilburg Clark, Lucius Beebe, and Charles Clegg, like their Bohemian counterparts, grew inexorably attached to Virginia City and the "Wild West" myth it embodied. In turn, the place became central to their lifestyles.

Tourist blight takes its toll

With the celebration of its centennial in 1959, the Comstock benefited from the emergence of a new TV show. Boasting 440 episodes, *Bonanza* took the country by storm, featuring beautiful color cinematography of Lake Tahoe and its environs, including a manufactured set purported to represent Virginia City. Unprecedented numbers of tourists flocked to the real-life Comstock to catch a glimpse of the American West fictionalized in weekly episodes. But the actual city looked nothing like the soundstage. So, some locals took the liberty of making the town over in the TV show's image. Changes included disguising finely-constructed nineteenth century brickwork under unfinished cedar-board facades that more closely resembled Hollywood false fronts. Michael Bedeau, Director of the Comstock History Center, refers to this bizarre architectural travesty as "Bonanza-fication," typified by the powder-blue exterior of the Bonanza Casino on C Street. By the mid-1960s, the extravagant influences of the tourist trade appeared all too visible.

In this sense, Virginia City represented a liminal space or interface between the nineteenth-century past and twentieth-century present (Turner 1977). To borrow a cliché, history literally came to life in this place. It was tangible, audible, visual, and even gustatory. The city seemingly thumbed its nose at the world and the changes wrought by time:

> Times might change, but the Comstock does not have to acknowledge that fact. The district could and will live on with its own brand of western and mining ethic, blithely ignoring its own evolution and the world around it, continually insisting that it maintains a firm anchor in the nineteenth century. Some aspects of the Comstock today seem clearly to be holdovers from its past. The occasional and unexpected opening of a deep nineteenth century shaft serves to remind everyone of why Virginia City was founded in the first place. Still, that is only one of the many echoes from its history.
>
> (James 1998, 273)

Despite the makeover, the wood-lined boardwalks, and haphazardly settled Victorian buildings lining C Street winked at twentieth-century tourists, only modestly eluding to the glory this crown jewel of the San Francisco mining period once enjoyed (Brechin 2001). The town clinched its fate as one of the great tourist destinations of the American West with an act of the U.S. Congress in 1961 designating Virginia City as a National Landmark. Those historic districts involved 400 buildings and some 14,500 acres of land. But the way in which Virginia City has emerged today is another matter altogether.

During the middle of the twentieth century, people flocked to Virginia City to gawk at cowboys, gunslingers, and Victorian ladies of the night. They did not come to see the impressive industrial machinery that the Comstock relied upon to access the rich ore deposits underneath its foundations. They came to experience the exaggerated, fantastical Wild West of Buffalo Bill, Wyatt Earp, Doc Holliday, and Jesse and Frank James. And they came in droves.

As the spotlight grew brighter on Virginia City, Beebe and Clegg grew weary of the dramatic rise in tourism. Conditions out of control with the premier of *Bonanza*, Beebe and Clegg mourned the loss of the city's character as it morphed into a twentieth-century tourist trap (Barnett 2014; James 1998). "Bonanza-fication" sickened the railroad buffs who had originally escaped to the area to experience the last palpable hints of America's frontier past. Socializing with middle-class tourists from suburbia had never been at the top of Beebe and Clegg's list. They simply could not stomach the transformation and retired to San Francisco. Other Comstockers who could afford the high cost of living joined them in San Francisco.

Beatnik and Hippie arrivals embrace and participate in the tourist frenzy

A generation later, another group of intellectuals, vagabonds, Beatniks, and proto-Hippies (or Hipsters) escaped to Virginia City (Curtis 2012). This time, however, the new arrivals embraced the city's tourist image becoming active participants in the tourist-induced fakelore. They reconfigured the Comstock House (an historic saloon) into one of the first stellar dens of psychedelic iniquity, the Red Dog Saloon (Barnett 2014; Works 1996) (Figure 10.1). Like their Bohemian predecessors, the Beatniks and Hipsters of the 1960s found an appeal in the area's remoteness and its general lack of law enforcement. Coupled with the generous tolerance afforded by locals and the predisposition to stay out of each other's business, twentieth-century Virginia City (and more specifically the Red Dog Saloon) functioned as a magnet for misfits, musicians, artists, and non-conformists.

Counterculture rock bands like the Charlatans and Big Brother and the Holding Company entertained Hipster crowds from across the Pacific Northwest and even as far afield as Michigan and Illinois. Hippie freaks like Ken Kesey and the Merry Pranksters made their pilgrimage to the Comstock, too. Place attachment ran strong for these counterculture misfits, and the area's current abundance of yard sculpture, hippie flags, tie-dye, and peace signs still attest to their permanent settlement in the area. According to long-time resident Mary Andreasen, "Virginia City was such a tolerant town then. There were many of those who espoused the [B]ohemian lifestyle, writers and artists, and that's the way it was" (Daley-Taylor 1992, 35). They were attracted by the Wild West images, seduced by the heady experience of unabashed self-expression, and quickly embraced a lifestyle that encouraged the performance and embodiment of the mythic West. Up in Virginia

Figure 10.1 The Red Dog Saloon on North C Street in Virginia City, Nevada.
Photo by author, 2014

City, the Red Doggers played out their wildest frontier fantasies (pulp fiction and film-western ethos) flavored with a dash of psychedelics (Barnett 2011, 2014, 2015; Palao 1996; Perry 1984; Perry and Miles 1997; Weller 2012; Works 1996). Chandler Laughlin, a kind of consigliere of the Red Dog Saloon, described the scene:

> We were the characters that sat around in the saloon down the street wait-ing for the boss to come and tell us to rustle the little girl's cattle so that she couldn't make the mortgage payment, you know. Everybody walked into the place and knew exactly where they were right away. Lynn Hughes' role was Miss Kitty in this thing. We were all doing all of the amalgamated Westerns that we'd grown up with in style including sitting in front of the place clean-ing our guns at dawn.
>
> (Works 1996, 0:52:12)

Many of the original Red Dog participants would become permanent fixtures of the community as they embraced the mythic lifestyle projected on the city (Laughlin 1995).

Returning to the land, connecting with a simpler time, and ignoring the world around appealed to Red Doggers. The retreat into fantasy and play rep-resented a means of psychologically dealing with the possibility of annihila-tion and impotence when faced with a government and "establishment" bent on projecting force and war. At the heart of all of this escapism and fantasy was not only the furtive beginnings of the back to the earth movement, but an intense longing to escape from a world of atomic bombs, communist plots, military draft, and disengaged communities. Taking on nineteenth-century per-sonas and "performing the Comstock" were essential features of their ultimate goal of finding something beyond what American mainstream culture wished to prescribe to them.

Through the Red Dog experience, they came to understand the "importance of setting down deep roots wherever [they] lived and forming a real relationship to the land itself." Moreover, they sought

> to "find the holy places" where [they] live[d] – the spring or grove or crest of a hill where [they] kn[e]w that others . . . lingered before [them] and steeped themselves, like [they], in its special stillness. Perhaps, though, the real point is not so much to *find* the holy places as to *make* them.
>
> (Flinders 1986, 30)

In this way, they reordered the arid landscapes of the Comstock holy places wherein they could unabashedly explore identity, meaning, and purpose. Virginia City and its environs, they felt, could transform their existence and validate their expanding consciousness, and today the original participants claim that this was, indeed, the case (Barnett 2011, 2014, 2015).

The Comstock as a validating place

Understanding Bohemian, Beatnik, and Hippie migration and place attachment to an unlikely Northern Nevada ghost town requires a rereading of essentialist conceptions of place to account for the fluidity of "multiple networked mobilities of capital, persons, objects, signs and information" (Light 2009, 241). Places are not simply fixed or bounded. They are not stable entities, but rather, perform their own kinds of transformation over time, and this, again, marked some of the brilliance of these twentieth-century subculture communities, who with not even one-tenth of the amenities and potentialities of the nineteenth century, produced a countercultural phenomenon still challenging the ways that we view the world, the environment, and our place therein today.

> In this context, tourism is one form of contemporary practice through which places and spaces are made and remade. Tourists do not simply encounter places, they also perform them: what a place is "like" depends, in part, on the nature of the embodied practices (enacted by both tourists and local people) that take place there.
>
> (Light 2009, 241)

Moreover, the performance of these expectations cement the bond humans feel with certain places (Lett 1983; Light 2009). Place attachment is ultimately as much about *why* people are attracted to certain places as *how* they express themselves in those places and embody their lifestyles of choice.

No matter their ultimate reason for coming, the new twentieth-century residents of Virginia City brought with them very specific expectations, and they actively attempted to fulfill these through fantasy, play, performance, and the preservation of folkloric remembrances. Even today, Virginia City's residents, especially those inhabiting and/or working on C Street, are well aware of the responsibilities of being a resident. They accept living in the "fish bowl" and play their unique parts (including daily donning costumes and fictional personas) with enthusiasm. They enjoy inculcating tourists with fables of the fictional past, although as I learned in interviews with diverse characters ranging from Pierce Powell to Squeak LaVake, they can relay accurate history, too.

By embodying place, history is rewritten and acted out daily on the Comstock. The performativity of this validating experience continues to profoundly impact the city's place myth underscoring carefully chosen elements while downplaying others (Baerenholdt *et al.* 2004; Light 2009). A strong place attachment persists and motivates surviving members of these avant-garde migrations to wage fierce opposition to open pit mining operations in the area. Ironically, they argue that mining threatens to destroy the historic sense of place that they have fabricated. Virginia City's colorful past attracted and continues to attract subculture adherents searching for validating space where they can freely explore escapism, role-playing, and fantasy. As avant-garde wannabe Cowboys, Bohemians, Beatniks, and Hippies employed the fakelore/folklore of the Comstock to inspire and reinforce personal and group identities. Today, the relationship remains reciprocal as they mold the city's cultural landscape to fit their expectations and beliefs.

References

Baerenholdt, J. O., M. Haldrup, J. Larsen, and J. Urry. 2004. *Performing Tourist Places*. Aldershot: Ashgate.

Barber, A. 2008. *Reno's Big Gamble: Image and Reputation in the Biggest Little City*. Lawrence: University Press of Kansas.

Barnett, E. 2011. Entertaining the Comstock. Pacifica. www.csus.edu/apcg/pacificaf11.pdf.

Barnett, E. 2014. Rockin' the Comstock: Exploring the Unlikely and Underappreciated Role of a Mid-Nineteenth Century Northern Nevada Ghost Town (Virginia City) in the Development of the 1960s Psychedelic Esthetic and "San Francisco Sound." Ph.D. dissertation, University of Nevada, Reno.

Barnett, E. 2015. The Comstock Summer of Love at 50. *Nevada Magazine*. http://nevada magazine.com/home/extras/the-comstock-summer-of-love-at-50/.

Beebe, L. and C. Clegg. 1949. *Virginia and Truckee: A Story of Virginia City and Comstock Times*. Oakland: Graham H. Hardy.

Brechin, G. 2001. *Imperial San Francisco: Urban Power, Earthly Ruin*. Berkeley: University of California Press.

Brown, G. and C. M. Raymond. 2007. The Relationship between Place Attachment and Landscape Values: Toward Mapping Place Attachment. *Applied Geography* 27: 89–111.

Browne, J. R. 1860–61. A Peep at Washoe. *Harper's Monthly* December 1860, January/ February 1861.

Cafferata, P. 2008. *More Than a Song and Dance: The Heyday of Piper's Opera House, Virginia City, Nevada, Circa 1863 to 1897*. Reno: Eastern Slopes Publisher.

Curtis, D. 2012. *Far Out: The University Art Scene, Reno, Nevada 1960–1975: An Exhibition and Conversation*. Reno: Special Collections, University of Nevada, Reno Libraries, July.

Daley-Taylor, A. 1992. Boardwalk Bons Vivants. *Nevada Magazine* 6: 20–24, 35–37.

DeLyser, D. 2005. *Ramona Memories: Tourism and the Shaping of Southern California*. Minneapolis: University of Minnesota Press.

DeQuille, D. 1974. *The Big Bonanza*. Las Vegas: Nevada Publications.

Dorson, R. M. 1977. *American Folklore*. Chicago: University of Chicago Press.

Flinders, C. 1986. The Work at Hand. In *The New Laurel's Kitchen*, edited by L. Robertson, C. Flinders, and B. Ruppenthal, 17–39. Berkeley: Ten Speed Press.

Florida, R. 2002. Bohemia and Economic Geography. *Journal of Economic Geography* 2: 55–71.

Gertz, S. J. 2010. *"Luscious" Lucius Beebe: Bon Vivant Book Man*. Seattle: PI.

Gross, M. J., C. Brien, and G. Brown. 2008. Examining the Dimensions of a Lifestyle Tourism Destination. *International Journal of Culture, Tourism and Hospitality Research* 2: 44–66.

Hausladen, G. 2003. How We Think about the West. In *Western Places, American Myths*, edited by G. Hausladen, 1–18. Reno: University of Nevada Press.

Hou, J. S., C. H. Lin, and D. B. Morais. 2005. Antecedents of Attachment to a Cultural Tourism Destination: The Case of Hakka and Non-Hakka Taiwanese Visitors to Pei-Pu, Taiwan. *Journal of Travel Research* 44: 221–233.

Hwang, S. N., C. Lee, and H. J. Chen. 2005. The Relationship Among Tourists' Involvement, Place Attachment and Interpretation Satisfaction in Taiwan's National Parks. *Tourism Management* 26: 143–156.

James, R. 1998. *The Roar and the Silence*. Reno: University of Nevada Press.

Kyle, G. T., A. J. Mowen, and M. Tarrant. 2004. Linking Place Preferences with Place Meaning: An Examination of the Relationship between Place Motivation and Place Attachment. *Journal of Environmental Psychology* 24: 439–454.

Laughlin, C. 1995. Rockin' at the Red Dog. *Nevada Magazine* July/August: 80–83.

Lett, J. W. 1983. Ludic and Liminoid Aspects of Charter Yacht Tourism in the Caribbean. *Annals of Tourism Research* 10: 35–56.

Light, D. 2009. Performing Transylvania: Tourism, Fantasy and Play in a Liminal Space. *Tourist Studies* 9: 240–258.

Lord, E. 1959. *Comstock Mining and Miners*. Berkeley: Howell-North Books.

Moehring, E. 1997. The Comstock Urban Network. *Pacific Historical Review* 66: 337–362.

Palao, A. 1996. *The Amazing Charlatans [CD Liner Notes]*. London: Big Beat Records.

Peck, G. 1993. Manly Gambles: The Politics of Risk on the Comstoc Lode, 1860-1880. *Journal of Social History* 26: 701–723.

Perry, C. 1984. *The Haight-Ashbury: A History*. New York: Rolling Stone Press.

Perry, C. and B. Miles. 1997. *Higher: The Psychedelic Era, 1965–1969*. San Francisco: Chronical Books.

Relph, E. 1976. *Place and Placelessness*. London: Pion Limited.

Rothman, H. K. 1998. *Devil's Bargains: Tourism in the Twentieth-Century American West*. Lawrence: University Press of Kansas.

Scannell, L. and R. Gifford. 2010. Defining Place Attachment: A Tripartite Organizing Framework. *Journal of Environmental Psychology* 30: 1–10.

Sheller, M. and J. Urry. 2004. *Tourism Mobilities: Places to Play, Places in Play*. London: Routledge.

Tsai, S. 2012. Place Attachment and Tourism Marketing: Investigating International Tourists in Singapore. *International Journal of Tourism Research* 14: 139–152.

Tuan, Y.-F. 1974. *Topophilia: A Study of Environmental Perception, Attitudes, and Values*. New York: Columbia University Press.

Turner, F. J. 1891. The Significance of History. *Wisconsin Journal of Education* 21: 230–336.

Turner, F. J. 1961. *Frontier and Section: Selected Essays of Frederick Jackson Turner*, edited by R. A. Billington, 1–37. Englewood Cliffs: Prentice Hall.

Turner, V. 1977. Variations on a Theme of Liminality. In *Secular Ritual*, edited by B. Myerhoff, 36–52. Amsterdam: Van Gorcum.

Twain, M. 2008. *Roughing It*. New York: Signet Classics.

Ullman, E. 1954. Amenities as a Factor in Regional Growth. *Geographical Review* 4: 119–132.

Watson, M. G. 1964. *Silver Theatre: Amusements of Nevada's Mining Frontier 1850 to 1864*. Glendale, CA: The Arthur H. Clark Company.

Weller, S. 2012. Suddenly That Summer. *Vanity Fair*, 14 June. www.vanityfair.com/culture/2012/07/lsd-drugs-summer-of-love-sixties.

Works, M. (Writer) and M. Works (Director). 1996. *The Life and Times of the Red Dog Saloon [Documentary]*. M. Works (Producer): Monterey Video/Sunset Home Visual Entertainment (SHE). 99 minutes.

11 Lost in time and space
The impact of place image on Pitcairn Island

Christine K. Johnson

Yi-Fu Tuan (1977) once wrote that, for a place to be *livable* it must display order and harmony between nature and society. Hidden away in a remote stretch of the southeast Pacific, roughly half way between New Zealand and Chile is a lush, tropical island that is home to 49 individuals. Pitcairn Island may be *livable*, but it is far from being a paradise. Pitcairn is infamously known as home to descendants of the HMAV *Bounty* mutineers. Over the past 227 years, the island has suffered devastating impacts to its place imagery. Yet, a small, cohesive population inhabits the island despite its turbulent past and poor reputation. This chapter traces the history of Pitcairn Island and examines the impact of Western media and ideals on its image. Despite tremendous odds, over 11 generations of residents on Pitcairn have developed an uncompromised attachment to their home. Not only is their identity derived from their unique culture, but also their shared struggle to overcome negative perceptions has strengthened their connection to the island.

I gathered data for this chapter from a variety of sources. I began by conducting a thorough review of secondary literature. Then I interviewed and directly communicated with a wide cross-section of island residents and official representatives. In 2004, I visited the island and conducted ten semi-structured interviews, and have since monitored publications and press releases pertaining to the island. I attended various Pacific-related conferences which provided additional information and allowed for direct communication with the Pitcairn Island Administration (PIA). Furthermore, I visited repositories in five nations that possess public and private collections of various forms and documents pertaining to Pitcairn Island. At these sites I reviewed historical records, special collections (i.e., photographs), and a variety of artifacts.

History of Pitcairn Island

The mutiny on the HMAV *Bounty* occurred on April 28, 1789, an event romanticized in film and literature for almost a century. As legend has it, a traumatized Fletcher Christian seized control of the British Navy's ship and dispatched her Captain, William Bligh, and his faithful followers to the *Bounty*'s cutter. Returning to Tahiti, the mutineers resupplied the ship, and set off seeking a hiding place capable of sustaining the nine British mutineers and their eighteen Polynesian companions (Alexander 2003). Christian knew Pitcairn to be mischarted, affording some degree of invisibility. The crew arrived on January 23, 1790 and, after verifying that the island was uninhabited and environmentally suitable, the mutineers set about the task of forming a community on the volcanic island. The thick, lush foliage concealed the community from the outside world for decades. Each year on January 23rd, Pitcairners celebrate the day the first settlers founded their community and culture. Residents of Pitcairn possess a strong sense of community and deep feelings of attachment to place rooted in their unique history. The settlement on Pitcairn endures today, despite repeated episodes of political, social, and environmental upheaval.

Because of its remote location, little research on Pitcairn Island has been done, and most of its history is derived from film and literature. The first known historical description of Pitcairn appears in the records of British explorer Philip Carteret, who passed by the island in 1767 and described it as "a great rock rising out of the sea, not more than five miles in circumference, and seem[ingly] uninhabited" (Carteret, quoted in Silverman 1967, 29). It is understandable why Fletcher Christian sought Pitcairn as a refuge; the island's isolation, verdant vegetation, and abundance of water were the key ingredients needed to sustain his mutinous crew.

Beginning in 1808, British Naval Captains of visiting ships provide the first accounts of the Pitcairn community. Early logs report that the inhabitants of the island were both civilized and devout (Mead 2007). For example,"[T]hey all speak English and have been educated by [Adams], the last living mutineer in a religious and moral way" (Folger, quoted in Alexander 2003, 351). The place served residents well, and the community thrived. The British Royal Naval Chronicle of 1816 described Pitcairn: "Small [as] Pitcairn's island may appear, there can be little doubt that it is capable of supporting many inhabitants, and the present stock being of so good [a] description, we trust they will not be neglected" (RNUK 1816, 21). However, this perspective of a happy, secure, and thriving community was to last just 15 years.

Place attachment in the face of resettlement

In the nineteenth century, the British government sought to resettle the Pitcairn population as a way to improve the quality of their life. Each attempt was met with stiff resistance. Although most of the earliest historical accounts describe the Pitcairn community as stable and content, by 1830 that perception changed. British authorities became concerned about drought and decided to remove the entire population (87 people). Despite expressed reluctance on behalf of the islanders, the British resettled the inhabitants on a "rich tract of land" in Tahiti granted by King Pomare (Shapiro 1936, 75). Soon after arriving, a number of Pitcairn natives were "taken sick," resulting in the death of 20 percent of the population. Those remaining returned to "their haven on Pitcairn" within six months. Captain Freemantle noted in 1833 that the Pitcairners affirmed their intentions to never leave Pitcairn again (Shapiro 1936).

In the mid-nineteenth century, British officials were once again concerned about the quality of life on Pitcairn Island. By 1849, the population had doubled, which prompted several ship captains to mention in their logs that space and resources on the island were becoming a concern. Local residents did not agree (Shapiro 1936). Perhaps the British government's concerns for the Pitcairners rested on the fact that they considered them fellow countrymen (Alexander 2003). Regardless, the British once again tried to relocate the island residents to what they perceived to be a more desirable location. In 1849, Captain Wood offered to transport the entire population off the island, and during the same year, Captain Fanshawe of the HMS Daphne reported: "I could not trace in any of them the slightest desire to remove elsewhere. On the contrary, they expressed the greatest repugnance to do so" (Fanshawe 1849, quoted in Shapiro 1936).

In 1856, the British Government yet again arranged for the islanders to be relocated, this time to Norfolk Island, approximately 1,000 miles northwest of Australia (Shapiro 1936). This island afforded much more space with each family being allocated 50 acres of land (Hinz and Howard 2006). After arriving, however, Norfolk wasn't the place the Pitcairners envisioned. The space was too vast, and in this way "uncomfortable" (Shapiro 1936). Shapiro noted:

> They missed the rugged beauty of Pitcairn and their cozy little houses embowered in rich foliage; they wanted the snug security that their own island gave them. The park like tranquility of Norfolk seemed immense. It lacked the dramatic beauty of their wild and romantic Pitcairn.
>
> (Shapiro 1936, 106)

It is evident that many Pitcairners had developed deep emotional ties to their small, remote island. By 1859, two families (16 people) left Norfolk and returned to Pitcairn. In 1864, another 27 people returned home (Shapiro 1936). It should be noted that only 25 percent of the original population returned to Pitcairn between 1859 and 1864 and the reasons for remaining in Norfolk have gone undocumented. Perhaps individuals' advanced age or the hardships of the month-long journey were factors that discouraged returning to Pitcairn.

Ultimately, every effort by the British government to resettle the entire population failed and after 1856 no further attempts have occurred. "Cultural identity for Pacific Islanders is most always tied to land" (Kahn 2011, 63). This idea is most evident by the actions of the nineteenth-century Pitcairners. After the last return to the island in 1864, life on Pitcairn returned to a peaceful and uneventful existence, and the Pitcairners were largely left alone for almost a century.

Perspectives of place

The physical geography of Pitcairn mimics many other islands across the Pacific Ocean (e.g., a high island with a subtropical climate) (Figure 11.1). As numerous Polynesian marketing schemes attest, these desirable qualities have long been affiliated with the concepts of paradise and island utopia. Unfortunately, Pitcairn

Island is not known for these traits. Rather, recent social issues and the infamous historical event regarding the *Bounty* endure.

The prevailing Western perspective focuses on the island as the site of the final resting place of the Bounty. Constructs by the West including media coverage, books, travel literature, films, and art focus primarily on largely negative imagery. The repeated reinterpretations, retellings, and representations of the mutiny on the Bounty saga undermines any positive imagery that the island projects. Pitcairn is painted as an unwelcoming place. As such, the present Pitcairn population suffers from the island's tarnished history.

Resident Pitcairners are moving forward to project a positive image of Pitcairn Island, one that focuses on the island's distinct culture and natural assets. Because the island is within Polynesia yet is maintained as an overseas territory of the United Kingdom, Pitcairn has a hybridized culture; the people are neither wholly British nor resolutely Polynesian. What this means is that with proper marketing, Pitcairn can capitalize on both its colonial history as well as its Polynesian indigenous background. Residents hope to build on this unique cultural foundation for tourism development. They see a potential in selling Polynesian material culture (e.g., baskets, honey, carved wooden curios) as well as ecotourism based on the natural environment (e.g., endemic species of flora and fauna, and the Pitcairn Islands Marine Reserve).

Figure 11.1 Pitcairn Island landscape.
Photo by author, 2004

Unfortunately, most efforts at tourism development have focused thus far on the history of the island, a history affiliated primarily with the British. This leaves Pitcairn Island seeming out of place within its geographic setting. The reality is that the founding population was predominantly Polynesian, living within Polynesia, yet choosing to affiliate with Britain for almost 20 years before British authorities knew they existed. Cultural connections to Britain include the incorporation of religion, language, and production of cultural material (e.g., architecture, clothing styles, furniture). Other elements of the Pitcairn culture stem from the Polynesian connections, seen in cooking methods, artwork, and other forms of cultural material (e.g., baskets, tapa). Additionally, the language of Pitkern (or Pitcairnese) was developed as an early method of communication; this pidgin language stemming from the blend of English and Tahitian languages is still spoken on Pitcairn and Norfolk islands today. Ultimately, the unique culture of the Pitcairners is largely overlooked, and the focus on British history, rather than the Polynesian geography, leads to a skewed perception of place.

Constructed images

From research published by anthropologist Harry Shapiro (1936) and journalist Ian Ball (1973), we know a considerable amount of information about modern Pitcairn culture. One thing that stands out is that with the exception of a few modern conveniences and a religious shift in the 1880s, little changed on Pitcairn during the late nineteenth and early twentieth centuries. Despite severe geographic isolation and limited tourism, the Pitcairn population remained strong, and life remained largely the same until the 1930s.

The perception that the outside world held for Pitcairn began to change dramatically after 1930. The shift started with the publication of the famous, but inaccurate trilogy of the *Mutiny on the Bounty* saga in 1934 by Charles Nordhoff and James Norman Hall. This was followed by the release of five major Hollywood films where producers liberally engaged in artistic license with respect to the mutiny legend. For example, Pitcairn Island has never hosted on-site filming of narrative films. Instead, for convenience, surrogate islands in the Pacific have served as stand-in. Furthermore, there is little consistency among the storylines represented in the five films, some focusing on character development (which is largely subjective), while others focus more on the paradisiacal island landscapes laden with romanticized (if not erroneous) images of the native inhabitants. Although producers may not have constructed these manifestations of place maliciously, they nevertheless have been less than faithful in their representation of the island and its settlers. From literature and film, the world community's perception of Pitcairn has been significantly distorted.

In 1983, a National Geographic article by Ed Howard critically describes Pitcairn as "a remote Pacific Island . . . a craggy, forbidding place . . . almost lost from the world. [It] loom[s] up like a tattered gray ruin of a fortress" (Howard 1983, 514–515). Clearly, he was not impressed with the physical setting. However, Howard continues in a more flattering tone when describing the cultural landscape:

The buildings are painted and orderly. The people of Pitcairn work almost continuously, gardening, cooking, getting firewood, mending something, carving curios, weaving baskets and hats, washing, sweeping, weeding the little cemetery, planting.

(Howard 1983, 520)

His description portrays a busy, engaged, and interactive community, comparable to positive descriptions of the quaint Polynesian culture of the eighteenth century (Figure 11.2).

In contrast, journalist Dea Birkett spent a year living on Pitcairn and wrote a scathing account of her experience in 1994. The book (*Serpent in Paradise*) depicts the darker side of utopia where "Pitcairn's single defining feature is its isolation" (Birkett 1997, 290). While stopping short of labeling Pitcairn a dystopia, the author's description leaves readers with the idea that the island is anything but paradise. The book was poorly received on the island, and had a significant negative effect on the perception outsiders have for Pitcairn and the people who live on the island.

As if the books and movies released during the late twentieth century did not do enough damage, the 2000s proved potentially catastrophic for Pitcairn. Four years into the new century, the tiny island became embroiled in the Pitcairn Sex Trials. Seven men were put on trial for crimes of a sexual nature. Sixty-four charges were filed, ranging from rape to sexual and indecent assault, dating back to 1964 (Marks 2008). Following lengthy appeals, six of the men were sentenced to prison on

Figure 11.2 Early Pitcairn community; residence of mutineer John Adams circa 1825.
Drawn and etched by Lt. Col. Robert Batty, 1831; published by Barrow in 1831

Pitcairn Island, and forced to build their own jail. Media coverage of the proceedings was extraordinary, given the spectacular nature of the crimes and the island's isolated location. What remained at the conclusion of the trials was a series of website postings, emails, blogs, magazine articles, and books, all of which painted Pitcairn as a shambles of a community housed in an isolated, dark, and desolate location. Phrases such as "Evil under the sun: The dark side of the Pitcairn Island," "Trouble in Paradise," and "Pitcairn Island, an Idyll Haunted by its Past," are just a few of the headlines that captured the event (Marks 2008; Allen 2013).

The media's descriptions of Pitcairn including, "craggy, menaced" (Allen 2013), "shattered, shocked, defiant" (Marks 2008), and "secret sex culture" (Prochnau and Parker 2008) run counter to the account provided by Sir Captain John Barrow in 1831. Using notes from a Captain Pipon who visited Pitcairn in 1814, Barrow declared, "that not one instance of debauchery or immoral conduct had occurred among these young people, since their settlement on the island" (Barrow 1831, 295). The modern descriptions and headlines cemented the image most people have of Pitcairn Island.

At the conclusion of the trials, journalists and politicians across the globe questioned the viability of the island. This time it was not concerns over drought or space to live on, but rather a concern for social justice. A 2008 article published in Vanity Fair quoted Pitcairn's new governor as saying "It is hard to exaggerate how damaging this is to those trying to persuade a wider world that Pitcairn's community can be sustainable" (Prochnau and Parker 2008, 139). Similarly, a 2008 memo from Britain's Minister for International Development stated, "The Pitcairn community is probably so socially dysfunctional that we should cease to plan to support and sustain it," adding that the Minister thought the best course of action was "voluntary depopulation" (Prochnau and Parker 2008, 137).

It is interesting to note that the prison sentences imposed upon the convicted men had a significant impact on the community. Because the total population of Pitcairn is less than 50 people, the removal of these men from the local workforce affected everyone. The prisoners represented a significant fraction of the resident wage earners, therefore their absence from the community was felt both physically and emotionally (Marks 2008). Furthermore, although no one condones their behavior, most people recognize that the trial will probably have a lasting impact on future population growth and tourism development of the island. It is also believed that this will be the final straw that encourages the British government to depopulate Pitcairn once and for all (Prochnau and Parker 2008). Today, Pitcairn seems to have lost the idyllic image once bestowed upon it. There is little positive imagery of the island outside publications that originate from the islanders themselves.

Placing history

"Toponyms are important for locating places within a landscape and allow for ease of spatial communication between two or more people" (Kostanski 2016, 424), and may perhaps be most perceptible in small or isolated places, such as islands. They are important on Pitcairn, not just for orientation. Rather, most of the place names are rooted in the history and culture of the island. The local toponyms represent important people and events as well as commemorate their *Bounty* origins. Pitcairn has a rich history and deep cultural traditions that are well-represented on the cultural landscape (Figure 11.3).

References to the origins and inception of the community as well as key ancestors are widely distributed on the landscape including place names like "McCoy's Valley," "Christian's Cave," "Brown's Water," "Matt's Rocks," and "Bounty Bay." Each of these can be categorized as commemorative toponyms, per geographer George R. Stewart's classification system (Howard 2016). "[N]ames have grown out of the life, and the lifeblood, of all those who had gone before" (Stewart 2008, 3). Most place names on Pitcairn range from association with exact coordinates to general areas, but all reflect some level of historical significance, and in this way further connect the islanders to the place.

As Laura Kostanski (2016) asserts, toponyms symbolize place attachment, as humans create a sense of place through the interactions with a place. Toponyms "identify the cultural mores of the communities that use the names," (Kostanski 2016, 416) where the Pitcairners themselves are part of the history, a living history tied to the land. Place identity is emotional in nature (Williams and Vaske 2003), and "place identity through identification with local histories can be similarly developed by both locals and non-locals" (Kostanski 2016, 417).

Figure 11.3 Sign post with location markers on Pitcairn Island.
Photo by author, 2004

Invested in place

The Tahitian proverb, "The life of the land is the life of the people," seems appropriate when considering Pitcairn's past. Irwin Altman and Setha Low (1992) define place attachment as the bonding of people to place (2). The actions of the Pitcairners historically represent this definition perfectly. As the source of more than half its founding population in 1790, Tahiti is most closely tied to Pitcairn's cultural history, and is, along with most other major island groups in Polynesia, still viewed in some capacity as a paradise and marketed as such. With the idea that place has a role in the construction of human identity and ties to land, the image Pitcairn Island bears today was externally constructed and is vastly different from what the Pitcairners have historically reflected (Korpela 1989; Dixon and Durrheim 2000; Kana'iaupuni and Malone 2010; Kahn 2011). Pitcairners have a broad range of emotions that connect to the place that is Pitcairn, but it is the positive experiences that create the attachment, the desire to maintain connection to a place and the senses of safety and comfort (Hernandez *et al.* 2007).

The islanders have been trying to maintain and market a landscape comparable to that of their Polynesian neighbors. The official website for Pitcairn Island (maintained by a resident of the island), offers detailed imagery and visitor information, and promotes the island's "environmental, economic and socio-cultural aspects of tourism" with a focus on "long-term sustainability" (Pitcairn Islands Tourism 2016). This, in spite of the notion that the island remains nearly as isolated today as it was 227 years ago when the mutineers and their companions first arrived. No air transportation is available, and all access is still reliant on passing or scheduled ships. As such, Pitcairn functions today much as a Western "gated community," defined as a settlement inhabited by socially homogenous residents surrounded by walls or fences (Blakely and Snyder 1999; Low 2004). However, the past decade has seen an active effort to bring back Pitcairners living off island.

The hope of increasing and rejuvenating the population is prevalent. There is no active desire by any of the adult population to abandon the island; rather, it is hoped by some that the younger generation presently being educated in New Zealand will return to reside on the island, despite the opinions and beliefs of particular British officials. Pitcairners continue to exhibit an emotional connection to their place based on their acknowledged and promoted history – their deep-rooted connection with this island, and their direct connection to their ancestors. The art of tapa has been rejuvenated, a skill practiced by the original matriarchs on the island, and perhaps most noteworthy, the land itself is part of their cultural heritage; the land registry book (land tenure in a paternal system) traces back to the earliest years of the nineteenth century, begun by the son of Fletcher Christian and describes the land holdings, brands, and other markings used to define and establish spaces on the island. While these markers are now gone or otherwise removed, the present land holdings are largely still associated with the earliest settlers. This added connection to the land is based in history, a history that largely defines the identity of the Pitcairners.

Pitcairn's tourism is presently based much more on the predominantly British history of the island, which leads to Pitcairn seeming out of place within its geography. While the indigenous populations of other Pacific islands have enhanced the sense of place and been a foundation for establishing tourism, it is Pitcairn's history to which most tourists are attracted. But unlike many Pacific islands that are facing environmental concerns, Pitcairn's sustainability is now measured by outside support. The island has recently been named home to the largest and most pristine marine environment in the world, officially established by the British Government and announced by British Prime Minister David Cameron in March, 2015 (Howard 2015), with a total area of 322,000 square miles (almost twice the size of the state of California). This added element of attraction may certainly increase tourism to the island, and is not rooted in the notorious history of the place. The present population is dedicated to presenting their island in the best light possible using the most modern technologies available. During the last decade, many of the island families have constructed apartments or extra rooms to house visitors, in anticipation of tourism growth, and with great sense of pride in their place.

"Place attachment contributes to individual, group, and cultural self-definition and integrity" (Altman and Low 2012, 4). As such, the Pitcairners are wholly individualized, with a unique and independent culture that is their own. However, infusion of the West through language, law, economic, and political control has impacted this island and its people throughout its modern history. "Image is about power. Those who can control the way a place is represented can control the place itself" (Borsay 2000, 7). Despite the many Western intrusions, there is a long-standing history demonstrating attachment to place by the islanders.

Past and future intertwined

Culture links members to place through shared historical experiences (Scannell and Gifford 2010). Over the years, with opportunities and resources available in other locales, some Pitcairners have emigrated, but retain deep-rooted ties to the land; some return late in life to live out their waning days on the island, some have returned posthumously to be buried there, and some of the younger generations have returned in the last decade, having temporarily sought education or employment off the island. However, the youngest generation (and smallest) generation of Pitcairners bears the greatest concern with regard to place attachment. There are just ten children in this newest generation, and most have been removed from the island beginning in 2011 for placement in New Zealand boarding schools to complete their education. In 1962, anthropologist Harry Shapiro added a postscript to his original 1936 publication about the Pitcairners. In this epilogue, he states:

> The educational opportunities for their children to be found in New Zealand, aside from obvious economic advantages, attract some of the families. . . . The fears of overpopulation under these circumstances need not trouble them any longer. Whether this interesting people will survive as a distinct population either on Pitcairn or elsewhere is open to question. Even if some islanders

remain on Pitcairn, their chances of survival become increasingly precarious as they are reduced in number, for the hazards to the survival of a small isolated population become proportionately great.

(Shapiro 1936, 256)

In this way, the Pitcairners face the same worry today as they did more than 60 years ago. However, life on Pitcairn has also changed significantly since Shapiro's time. Residents are now better connected to the outside world through television, internet, and modern cell phones. These amenities help overcome the fact that traveling to the island remains difficult. However, better digital connectivity does not counteract the strong pull exuded by the education and employment opportunities found in New Zealand (and other countries). The world of 2016 is a far different place than even Shapiro could imagine in 1936. In an interview in 2004, a Pitcairner of the oldest generation reflected: "I worry that if the children leave, we will forget our history; we will forget the reasons behind the place names and the places will be lost forever."

The identity of the Pitcairn population stems from a single historic event that occurred in 1789. The resulting establishment of a community living in isolation as a result of mutiny resonates centuries later, as we watch this unique population strive to exist in the face of multiple challenges. The Pitcairn cultural identity is an interesting blend of the British and Polynesian cultures, and enhanced by today's political ties to Britain and geographic situation within Polynesia. Most residents share common heritage, proudly tracing their ancestry to the original settlers, and have a proven attachment to the place, returning multiple times after being removed. The common thread holding all of it together is shared struggle, surviving together in the face of outside threats and negative perceptions.

The Pitcairn population lives a precarious existence – small, remote, economically marginal – yet have thrived for over 200 years. Part of their identity is derived from struggling to live on that island. Despite all the infamous history and negative publicity, the Pitcairn population has remained steadfast and rooted in place. There are many obstacles on the horizon (youth leaving, depopulation, economic vitality in the modern age, competition in tourism), but the people of Pitcairn are more than just hanging on. They are deeply tied to the place, with cultural and historical details that help validate their lives.

References

Alexander, C. 2003. *The Bounty: The True Story of the Mutiny on the Bounty*. New York: Viking.

Allen, K. Pitcairn Island, an Idyll Haunted by its Past, Star. 16 December 2013. www.thestar.com/news/world/2013/12/16/pitcairn_island_an_idyll_haunted_by_its_past.print.html. Last accessed 23 February 2014.

Altman, I. and S. M. Low. 1992. *Place Attachment*. New York: Plenum Publishing Corporation.

Altman, I. and S. M. Low. 2012. *Place Attachment*. Berlin: Springer Science & Business Media.

Ball, I. 1973. *Pitcairn: Children of Mutiny*. Boston: Little, Brown & Company.

Barrow, J. 1831. *The Eventful History of the Mutiny and Piratical Seizure of H.M.S. Bounty: Its Causes and Consequences*. London: John Murray.

Birkett, D. 1997. *Serpent in Paradise*. New York: Anchor Books.

Blakely, E. J. and M. G. Snyder. 1999. *Fortress America: Gated Communities in the United States*. Washington, DC: Brookings Institution Press.

Borsay, P. 2000. *The Image of Georgian Bath, 1700–2000*. New York: Oxford University Press.

Dixon, J. and K. Durrheim. 2000. Displacing Place-Identity: A Discursive Approach to Locating Self and Other. *British Journal of Social Psychology* 39: 27–44.

Hernandez, B., C. M. Hidalgo, M. E. Salazar-Laplace, and S. Hess. 2007. Place Attachment and Place Identity in Native and Non-Natives. *Journal of Environmental Psychology* 27: 310–319.

Hinz, E. R. and J. Howard. 2006. *Landfalls of Paradise: Cruising Guide to the Pacific Islands*. Honolulu: University of Hawai 'i Press.

Howard, B. C. 2015. World's Largest Single Marine Reserve Created in Pacific. National Geographic. http://news.nationalgeographic.com/2015/03/150318-pitcairn-marine-reserve-protected-area-ocean-conservation/. Last accessed 11 January 2017.

Howard, E. 1983. Pitcairn and Norfolk: The Saga of Bounty's Children. *National Geographic Magazine* October: 510–529.

Howard, P. 2016. *The Ashgate Research Companion to Heritage and Identity*. New York: Routledge.

Kahn, M. 2011. *Tahiti beyond the Postcard: Power, Place, and Everyday Life*. Seattle: University of Washington Press.

Kana'iaupuni, S. M. and N. Malone. 2010. This Is My Land: The Role of Place in Native Hawai'ian Identity. In *Race, Ethnicity & Place in a Changing America*, edited by J. W. Frazier and E. L. Tetley-Flo, 291–305. Albany: State University of New York Press.

Korpela, K. M. 1989. Place-Identity as a Product of Environmental Self – Regulation. *Journal of Environmental Psychology* 9: 241–256.

Kostanski, L. 2016. *The Oxford Handbook of Names and Naming*, edited by C. Hough and D. Izdebska, 416–418, 424–425. Oxford: Oxford University Press.

Low, S. 2004. *Behind the Gates: Life, Security, and the Pursuit of Happiness in Fortress America*. New York: Routledge.

Marks, K. 2008. *Pitcairn - Paradise Lost: Uncovering the Dark Secrets of a South Pacific Fantasy Island*. New York: Harper Collins.

Mead, J. K. 2007. National Register, Volumes 800–801 of American Periodical Series, Volume 1, Issue 1 and Volume 2, Issue 43. Harvard University, 1816. Books.google.com. Last accessed 29 June 2013.

Nordhoff, C. and J. N. Hall. 1934. *Mutiny on the Bounty*. Boston: Little, Brown & Company.

Pipon, P. 1814. *Captain Pipon's Narrative of the Late Mutineers of H.M. Ship Bounty Settled on Pitcairn's Island in the South Seas, in Sept, 1814*. London: Mitchell Library, Sydney Manuscripts: Banks Papers.

Pitcairn Islands Tourism. 2016. www.visitpitcairn.pn/pitcairn/general_info/index.html. Last accessed 25 February 2014.

Prochnau, W. and L. Parker. 2008. Trouble in Paradise. *Vanity Fair* January: 95–103, 136–139.

Royal Navy of the United Kingdom (RNUK). 1816. *Naval Chronicle for 1816: Containing a General and Biographical History of the Royal Navy of the United Kingdom: With a*

Variety of Original Papers on Nautical Subjects under the Guidance of Several Literary and Professional Men. Volume 35, January to June. London: Joyce Gold.

Scannell, L. and R. Gifford. 2010. Defining Place Attachment: A Tripartite Organizing Framework. *Journal of Environmental Psychology* 30: 1.

Shapiro, H. L. 1936. *The Heritage of the Bounty: The Story of Pitcairn through Six Generations*. New York: Simon and Schuster.

Silverman, D. 1967. *Pitcairn Island*. Cleveland: World Publishing Company.

Stewart, G. R. 2008. *Names on the Land: A Historical Account of Place-Naming in the United States*. New York: New York Review Books.

Tuan, Y.-F. 1977. *Space and Place: The Perspective of Experience*. Minneapolis: University of Minnesota Press.

Williams, D. R. and J. J. Vaske. 2003. Measurement of Place Attachment: Validity and Generalizability of a Psychometric Approach. *Forest Science* 49: 830–840.

Part VI

Vanishing places

12 Rethinking Fountainbridge

Honoring the past and greening the future

Geoffrey L. Buckley

If you have been a tourist in Edinburgh, Scotland you are likely familiar with many of the city's historic landmarks. You may have strolled the ramparts of its famous castle, toured the Palace of Holyroodhouse, or explored the Old Town's medieval closes and wynds. Perhaps you climbed to the top of the Scott Monument, Calton Hill, or Arthur's Seat, to gain a different perspective on this ancient capital. Maybe you even signed up for a ghost tour that introduced you to the city's "haunted" underground chambers and vaults. If it was souvenirs or gifts you desired, doubtless you investigated shops along the famous Royal Mile, many of which peddle highly romanticized images of Scotland's history to appeal to the imaginations (and pocketbooks) of sightseers. Maybe you dined in the New Town, on George or Rose Street, admiring the Georgian architecture and the symmetry of James Craig's 1765 plan. Chances are you did *not* pay a visit to Fountainbridge.

Located adjacent to the World Heritage Site in Edinburgh, the Fountainbridge neighborhood has its own storied past. In the nineteenth century, the North British Rubber Company (NBRC) employed thousands of workers to churn out a wide range of consumer goods, from rubber galoshes and "wellies" to rubber tires and hot water bottles. During World War I and World War II, the children and grandchildren of many of these same workers produced boots, gas masks, and other materiel for export both home and abroad. Other industries left their imprint on the neighborhood as well, including printing and stationery, canning, textile manufacturing, meatpacking, and dairying. Before achieving fame and fortune on the big screen in the 1950s, Sean Connery (arguably Fountainbridge's most famous export) got his start driving a milk cart for the St. Cuthbert's Cooperative Society. Breweries and distilleries were also important employers. An important feature of the neighborhood was, and remains, the Union Canal, which connected Edinburgh with Glasgow and served as a conduit for the export of goods for decades. Of course, this is not the history most vacationers have in mind when they visit Edinburgh. Nor is it the sort of history city officials seek to promote.

Outside the boundaries of the World Heritage Site, the Edinburgh Council and the city's various tourism agencies are considerably less interested in the history that places like Fountainbridge have to offer, especially if it has anything to do with industrial heritage. In Fountainbridge and elsewhere, local authorities in recent years have approved plans to raze entire city blocks of working-class tenements and industrial structures, replacing them with steel and glass office buildings, apartments, and student accommodations. For many former residents, the transformation has been disorienting. According to the *Daily Record and Sunday Mail* from 18 June 2010, a "baffled" Sir Sean Connery said he felt "lost" as he toured the Edinburgh streets where he was born and bred. Back in town to dedicate a plaque marking the site of his birthplace (the actual tenement where he was raised had long since been knocked down) "the Bond legend hardly recognised the redeveloped area of Fountainbridge that used to be his stomping ground . . . Fresh from his red-carpet appearance at the Edinburgh International Film Festival," the 79-year-old Connery "admitted he hardly recognised Fountainbridge," commenting to the press "I'm lost walking round here" (*Daily Record* 2010).

Since the economic downturn of the late 2000s, however, local activists have taken advantage of a lull in demolition and construction to campaign for an alternative vision for their neighborhood. Formed in 2011, the Fountainbridge Canalside Initiative (FCI) has been working with developers and others, to honor the area's industrial past, while at the same time, plan for a more just, livable, and sustainable future. With respect to green space (always in short supply in Fountainbridge) residents and their allies have embraced a truly novel approach to urban greening that holds promise for other cities in Europe and the United States faced with similar brownfield redevelopment challenges.

In this chapter I make four key points. First, post-World War II planning efforts have sought to transform Edinburgh from a center of industrial activity to a cultural capital (a view that is supported by the existing literature). Second, while World Heritage Site designation in 1995 called for protecting the Old and New Towns, it intensified development in other neighborhoods such as Fountainbridge. As locals know all too well, however, even a site located deep in the heart of the Old Town is not guaranteed protection, as the recent controversy over a proposed development scheme in the Cowgate plainly shows (*The Guardian* 2016). Third, since 2011 FCI has given the local community a voice in the planning process. Already, it has successfully protected the last historically significant "listed" building on the old rubber company property, lobbied for the construction of a new high school, secured height limitations on new buildings, and launched two mobile community gardens. Finally, FCI and its partners have offered residents an alternative way of engaging history, one that resonates with the local community and their attachment to place.

As Benton-Short (2006), Dwyer and Alderman (2008), Foote and Grider (2010), Alderman and Dwyer (2012), Alderman and Inwood (2013), MacDonald (2013), Post and Alderman 2014), Foote (2016), Post (2016), and numerous others have shown over the years, how we commemorate the past (and who gets to tell a given story) matters. Likewise, scholars of environmental psychology including Proshansky *et al.* (1983), Mesch and Manor (1998), Vaske and Kobrin (2001), Brown *et al.* (2003), and Manzo and Perkins (2006) remind us that place attachment is an essential element when it comes to building community and fostering environmental stewardship. While no one is arguing that the City of Edinburgh should not have targeted the dilapidated industrial brownfield at the heart of the Fountainbridge neighborhood for preservation, it is clear that the site was historically significant and that plans for its redevelopment should have taken into account the needs and desires of local residents from the outset. Generations of workers have lived and worked in Fountainbridge, and over that time they have developed a strong attachment to place. Even more recent arrivals to the community have established a bond to the area's industrial past. It is imperative that this history not be completely effaced to promote a more tourist-friendly and sanitized history.

Edinburgh's industrial past

Throughout the nineteenth and much of the twentieth century, Edinburgh's economy was much more dependent on industry than it is today. As late as 1951, 50.7 percent of the male workforce and 37.7 percent of the female workforce were employed in the industrial sector where they engaged in a wide range of activities. These included machine and shipbuilding; foundry and metalworking; brick, cement, and glass manufacture; milling and food processing; pottery and chemicals; textiles and clothing; and optical and precision instruments (Madgin and Rodger 2013, 511). Such diversification distinguished Scotland's capital from other cities in the United Kingdom more closely aligned with a single industry. It also insulated Edinburgh's workforce from the "boom and bust" fluctuations that wreaked havoc elsewhere (Rodger 2005).

At the same time that Edinburgh's industrial sector was expanding between 1840 and 1910, "Victorian observers" were downplaying its prominence as a manufacturing city. Instead they promoted an image that highlighted its historic and picturesque features (Rodger 2005; Madgin and Rodger 2013). In their detailed examination of the city's living conditions ca. 1865, Laxton and Rodger (2013) paint a picture that contrasts sharply with the one envisaged by Edinburgh's *bourgeoisie*. It was an "insanitary city," one that in the eyes of medical officer Henry Duncan Littlejohn "extended beyond inadequate drainage, sewers, and water supply to include poor amenity and intolerably unpleasant environment" (Laxton and Rodger 2013, 125). To Littlejohn, industrial developments such as the new Caledonian Distillery chimney in Dalry, not far from Fountainbridge, were not just eyesores "but by intruding into residential areas they injured health and depreciated neighborhoods" (Laxton and Rodger 2013, 126). Industry's influence was everywhere to be found. Nevertheless, despite an abundance of evidence to the contrary, the "myth" of non-industrial Edinburgh persisted well into the twentieth century.

As World War II drew to a close, city officials and town planners across Great Britain turned to the task of rebuilding. More than five years of German bombing raids had inflicted heavy damage on important industrial and administrative centers like Birmingham, Coventry, Hull, London, and Plymouth. Destruction was so extensive in some districts that planners were essentially working with a blank slate. Perhaps because Edinburgh was not dominated by any one industry, such as Clyde River shipbuilding in Glasgow, it emerged from the war largely unscathed. Surprisingly, this did not deter planners from focusing attention on Scotland's capital. Rather, they viewed it as an opportunity to reflect on Edinburgh's past, evaluate its present, and rethink its future. According to one prominent report: "The authorities whose cities had escaped devastation rightly realized that the inherent problems were similar to those of the bombed towns" (Abercrombie and

Plumstead 1949, vii). After all, these cities too had to contend with housing problems, overcrowding, and industrial pollution.

In Edinburgh, rather than continue to understate the importance of industry, as had been the case for so long, politicians and planners now targeted industrial areas for renewal. As Rebecca Madgin and Richard Rodger (2013, 518) aptly put it: "After decades of denial, industry could no longer be ignored." Indeed, for planner Patrick Abercrombie and his assistant Derek Plumstead, industry *was* the problem (Morris 2010). Starting in the late 1940s and then gaining momentum thereafter, Edinburgh moved aggressively to construct an image of itself more befitting of a cultural capital, withdrawing space for industrial use while promoting policies that favored office, retail, culture, and tourism (Madgin and Rodger 2013). Nowhere is this more evident than in Edinburgh's Old Town. Here, politicians, planners, and architects promote a vision of the city that dovetails with what tourists have come to expect when they visit Edinburgh. Even new construction conforms to the Scottish Baronial style, an invention of the nineteenth century that has become synonymous with the city's historical narrative. Walking the streets of Old Town, it is difficult but still possible to find traces of the city's industrial past, remains that have been repurposed so that they are compatible with an economy that now caters overwhelmingly to tourists. Similarly, it is relatively easy to identify buildings or parts of buildings that have been constructed since the 1940s to fit the image of what people anticipate when they visit a "medieval" city. Upon closer inspection, even "ancient" historic landmarks like Edinburgh Castle are, in part at least, inventions of the twentieth century (Morris 2007).

In 1995 the United Nations Educational, Scientific, and Cultural Organization (UNESCO) added the Old and New Towns of Edinburgh to its list of World Heritage Sites. As the city's management plan indicates, Edinburgh's unique planning history (see McKean 2005) and efforts at historic preservation figured prominently in the organization's decision:

> The remarkable juxtaposition of two clearly articulated urban planning phenomena. The contrast between the organic medieval Old Town and the planned Georgian New Town provides a clarity of urban structure unrivalled in Europe. The juxtaposition of these two distinctive townscapes, each of exceptional historic and architectural interest, which are linked across the landscape divide, the "great arena" of Sir Walter Scott's Waverley Valley, by the urban viaduct, North Bridge, and by the Mound, creates the outstanding urban landscape.
>
> (World Heritage Site Management Plan 2011–2016, 14)

While the creation of a World Heritage Site in Edinburgh protected "townscapes . . . of exceptional historic and architectural interest," it also deflected new development to other locales. For communities like Fountainbridge and Craigmillar (another industrial hub that has been all but erased from the map), the long-term consequences have been significant (Kallin and Slater 2014). Of course, history played out in these places too, but, in the eyes of planners and developers it was the wrong kind of history.

Residents react

Not long after World Heritage Site designation was conferred on Edinburgh's Old and New Towns, a significant portion of Fountainbridge's industrial past was swept away. In February 2004, Scottish and Newcastle (S&N) announced it was shuttering its brewery in Fountainbridge as a cost-saving measure, bringing to a close 148 years of brewing the famous McEwan's brand on site. According to a report in *The Scotsman* (2010), "Throughout the 19th century, and up until the 1960s, brewing was a massive industry in the capital. At one point the Fountain Brewery was just one of 30 in the city, but S&N's closure decision left just one in the city, the nearby Caledonian Brewery."

After the brewery closed in 2005, plans were finalized to demolish the site. By 2011, the city cleared a total of 13 acres including factory buildings, tenement housing, and other structures, leaving behind a desolate and derelict landscape (Figure 12.1). The exception was one "grade-2" listed building, the old headquarters of the North British Rubber Company. Then city authorities announced that it, too, would be knocked down. The demolition was part of a wider regeneration program, approved by city council that called for luxury flats, student accommodations, short-term rentals, and office complexes. Beginning with the canal redevelopment project called Edinburgh Quay, the new owners of the land (Grosvenor, Lloyd's Banking Group, and Edinburgh City Council) got down to the business of redeveloping the massive site.

Interviews conducted with long-time residents reveal a sense of frustration and loss. Not only were they left out of the decision-making process, they lamented

Figure 12.1 Construction fence along Fountainbridge Road, facing northwest. Barriers such as this abound in Fountainbridge. They isolate residents, obstruct sightlines, and disrupt pedestrian flow. Note the top of the smokestack at Dalry in the distance (one of the few reminders of Edinburgh's industrial past).

Photo by Ingrid Buckley, 2014

the loss of a place (albeit now an abandoned one) to which they had developed a strong attachment. According to one female informant:

> My earliest memories of Fountainbridge go back to 1969 when I first came to the city. In the 1970s I lived just behind Haymarket Station in a row of condemned Victorian railway workers cottages, now demolished, and I would enter Fountainbridge from there to access the Meadows, Tollcross, and the art college on Lauriston Place. It was a transit route for me, never a destination or a place to visit on its own, but there were still some older buildings along Dundee Street/Fountainbridge near Grove Street which were memorable and intrigued me. . . . The clock was a notable and welcomed landmark. It and the older buildings suggested a time when the area once had a dignified human identity.

Other residents recalled a time when the brewery was a fixture in the neighborhood. A long-time male resident stated:

> The whole area was dominated by the brewery. . . You could smell the malt from the brewery quite strongly. The smell was as strong a part of my memory of the impression of the place. You could smell it down at Haymarket and up at Tollcross. It would even drift up to Polwarth and Bruntsfield on occasions.

When the facility closed, the landscape changed: One woman remembers that,

> The land covered by the brewery dominated and it was a sad, bleak, impersonal, scary, dead, industrial area. The older buildings that I liked often felt fragile, vulnerable, like a frail but venerable and dignified old lady, clinging to life against all the odds.

Residents complained about historical marginalization and the City Master Plan's emphasis on high-end development. According to one retired male resident: "The first thing they did was to rename the place Edinburgh Quay. It's Lochrin Basin. You rename a place and you destroy it." The plan was "a complete negation of the culture and the place," he continued. "Luxury flats and Grade 1 offices in the midst of a working class neighborhood. None of the people that lived here would be able to go to these places." Also, many nearby residents did not like the fact that their new neighbors would be college students and short-term renters, populations unlikely to invest in the community or participate in community affairs. Others associated the decline of Fountainbridge with the World Heritage Site label: "Three blocks that way it's all about preservation," remarked another retired male resident. "Here it's anything goes." Advertisements posted throughout the site suggested a market audience younger, wealthier, and more transient than the current resident population. Featuring pictures of college-aged men and women, they tout the amenities of new and planned buildings: "Free Gym and Social Spaces,"

"Hotel Style, Home Comfort," "Grade A Offices Available To Let." One banner with the optimistic phrase "Perfection is a matter of time" emblazoned on it beckoned customers to a new short-term rental property.

The Fountainbridge Canalside Initiative

Then, in the late 2000s, the world economic crisis caused construction to grind to a halt. The respite from building gave concerned citizens the opportunity and time needed to organize and strategize. What ultimately emerged from these preliminary meetings was the Fountainbridge Canalside Initiative (FCI), an advocacy group that gave the local community a voice in the planning process. Officially established in August 2011, the group hosted its first major public meeting (an event called "Brewing New Life") in February 2012. Billed as a "visioning" exercise, the occasion attracted more than 80 people including local politicians and representatives from a broad cross-section of community organizations. FCI described the event as "a lively day of discussions and proposals for the future development of the canalside sites" (Fountainbridge Canalside Initiative 2012a). The purpose of the meeting was fivefold: 1) to "generate discussion, support and ideas," 2) determine "what local people and groups thought most important," 3) establish priorities for the group, 4) attract new members and contacts, and 5) secure "a strong mandate from the community to move FCI forward" (Reeves 2014).

Over the course of the next five years, FCI members worked effectively with city officials and developers to modify the master plan and influence new development, emphasizing their concern for social justice and place making (Fountainbridge Canalside Initiative 2012b). For instance, they advocated for height limits on new construction and were successful in their efforts to curtail the number of student accommodations built in the neighborhood. They also supported the relocation of the Boroughmuir School to the western end of the construction site, mixed-income housing, and the addition of a new green space to which the general public would have access (Fountainbridge Canalside Initiative 2012a, b, c, d, 2013, 2014). With respect to historic preservation, FCI and its allies were able to protect the North British Rubber Company headquarters from demolition and, eventually, find a new tenant to restore and occupy the building (Fountainbridge Canalside Initiative 2012a, b, d, 2013, 2014) (Figure 12.2). Significantly, FCI received approval to introduce "mobile gardens" to vacant sections of the site. Perhaps more than any other initiative, the development of the Grove 1 and Grove 2 community gardens galvanized support for FCI and instilled a sense of ownership and community spirit.

The motivation for creating a mobile community garden in Fountainbridge was that it could serve as a "meanwhile project" on vacant land where development was stalled. Hoping to take advantage of technical expertise from an organization in Glasgow called Sow and Grow Everywhere (SAGE), FCI sought out support from willing developers. When FCI learned that funding for SAGE had unexpectedly run out in Fall 2012, members of FCI and their partners were compelled to form a new organization in order to negotiate a leasing agreement for 1,600 square meters with Grosvenor Development (the principal developer involved in the project). Drawing on a mailing list of 90 people, the new group (dubbed The Grove) was organized with FCI serving as parent organization (Fountainbridge Canalside Initiative 2012d, 2013). While the SAGE model called for soil and growing medium to be placed into one-ton "builders bags" that could be transported to other sites, members of The Grove opted instead to use recycled planters and pallets as a more cost-effective option (Fountainbridge Canalside Initiative 2012a).

Landscape gardeners employed by Grosvenor then prepared the site for gardening, fenced it in, and connected all the services. Forty plots were allocated to named gardeners, with 35 situated on the periphery to serve as communal plots. As one founding FCI member put it: "This mix of individual's responsibility, and collective caring for the site, and the communal plots, is one of the key ideas and strengths of the project" (Reeves 2014). The gardeners then got down to work, turning over their plots, adding compost donated by the City Council, installing a green house, and planting seeds and plants. "In the exciting first weeks," recalls Reeves (2014), "much sharing, help and advice went on, and by the middle of June the whole site was blooming." Enthusiasm for gardening notwithstanding, organizers recognized the need to train novice gardeners. Members of The Grove,

Figure 12.2 Grove 2 Community Garden. The use of wooden planters and pallets will facilitate removal to a new site as development proceeds. The North British Rubber Company (NBRC) Headquarters (with boarded up windows) is visible in the middle background. To the left of the NBRC building is an old tenement and beyond that is the new Edinburgh Quay development.

Photo by Ingrid Buckley, 2014

therefore, began to convene "workshops, learning events, and informal exchanges over a cup of tea at the garden" (Reeves 2014).

In addition to providing fresh fruits, vegetables, and flowers, the Grove Community Garden has proven popular for social reasons. Testimonials from gardeners suggest the garden has helped build a sense of community in an area devastated by demolition. One female resident stated: "The garden makes me feel included in a group of people outside work and other stuff I do. I would never have met all these different kinds of people and learn so much about gardening in such short time" (quoted in Reeves 2014). Another notes that, "Gardening is usually a solitary affair but this is different. It's great to have somewhere to go where you know it will be peaceful. I love that there's such a mix of people to get to know" (quoted in Reeves 2014). According to Reeves (2014), "The Grove stands as an example of humanized urban development, which could be replicated in many cities, growing nature and community." In February 2014, The Grove entered a lease agreement for a second plot – the Grove 2 Community Garden – that opened the following month (The Grove Community Garden 2014).

Another important aspect of FCI's mission is to protect what remains of the community's industrial heritage. A founding member of FCI put it this way:

> [W]e feel strongly that the bleakness and alienation of the former brewery as an industrial site wedged between residential areas should not be replaced by the anomie of student housing, hotels, and offices, but with something living that builds on the historic identity of Fountainbridge before the brewery came. This historic identity is still faintly perceptible and we hope it can be built on.

She added: "There was a strong sense of community in this matrix at whose centre was NBRC with its many social spin offs, clubs, and sporting societies." To this end, FCI members "want the history of the area to be reflected in its future," by preserving the canal quayside, installing a sculpture in the square "to remember the important contribution the rubber works made to Edinburgh's industrial heritage," and the erection of a community facility that will "provide space for contemporary industrial innovation." FCI would also like to see the famous McEwans clock from the brewery "restored and integrated in a prominent position" (Fountainbridge Canalside Initiative 2015a).

To date the city's attempts at historic preservation have been more modest. Instead of saving structures, officials have installed plaques and other forms of memorial to acknowledge the area's history. Two memorials, in particular, bear mentioning. The first is located in a quiet courtyard next to a new office building. A large photograph reproduced from the pages of a 29 January 1937 edition of *The Evening Dispatch* shows animals being driven to slaughter in what was once the meatpacking center of the city. "Sheep's plucks and bags. Come awa'" is inscribed three times across a section of the image. To the left is a passage from the newspaper that describes the conditions of Fountainbridge in the 1870s:

> Fountainbridge of sixty years ago had a much different aspect from that of the present day [1937]. The thoroughfare was narrow and congested, and dirty and

greasy from the numerous droves of sheep and bullocks being driven to and from the Cattle Markets at Lauriston or to the slaughter-houses, which then occupied the site of the recent school between the Palais de Danse and Ponton Street, which then mostly comprised small, red-tiled, white-washed houses.

(*The Evening Dispatch* 1937)

The passage goes on to describe workers "laden with the raw ingredients for haggis making [a pudding made from sheeps hearts, livers, and lungs]" who "seemed always to be doing a roaring trade." It was described as a "busy place at all times" with laborers "unloading the barges of coal, bricks, fireclay, and stores at the different wharves." While the description and photograph are evocative of the time period, today the site is hidden and relatively sterile.

The second memorial commemorates the construction of the Union Canal. Embedded in the ground at the edge of a wide sidewalk, the marker (about the size of a sidewalk paver) is small and inconspicuous. An advertisement for a local café is often placed next to it, obscuring its presence from passersby. The text associated with the marker describes major events in the canal's history, including several weeks during the winter of 1878–79 when it froze completely. A miniscule map shows the former terminus of the canal at Port Hamilton and Port Hopetoun, locations that have since been filled. As with the memorial remembering the meat-packing industry, today the canal marker appears to go largely unnoticed.

In contrast to the city's attempts to "tell the story" of Fountainbridge (displays that residents argue are concealed from view and ignored by the general public), FCI and its partner organizations enlisted community members to tell their own stories about Fountainbridge. The "wall panels" they produced describe the labor history of the area, from the digging and operation of the canal to the many trades that located here, including brewing, silk shawl weaving, and numerous other artisanal industries (Fountainbridge Canalside Initiative 2015a). The panels are exhibited at various events throughout the year, but most prominently at the annual canal festival, which celebrates the neighborhood's past while providing residents with an opportunity to shape its future (Figure 12.3). Also on display at the canal festival are blueprints of the neighborhood master plan. Residents and visitors are invited to study the plan, review proposed changes, offer opinions, and learn more about the work of FCI. Visitors are also encouraged to tour the Grove community gardens as well as other meaningful attractions such as The Forge woodworking and metalworking open access workshops. Most important, the Canal Festival is a place to have fun in Fountainbridge as evidenced by the crowds that attend the event's popular raft race.

FCI unveiled its most recent achievement in October 2015 when more than 70 volunteers assembled a "standalone community space" (the Fountainbridge Community Wikihouse) in the heart of "a prominent brownfield plot" owned by the City of Edinburgh Council (Fountainbridge Canaside Initiative 2015b). Led by members Jane Jones and Akiko Kobayashi, the group acquired the funding and resources necessary to construct a shelter that hosts workshops and events focusing on "locally relevant issues" such as community empowerment, housing, land rights, and participatory urbanism (Fountainbridge Canalside Initiative 2015b).

Figure 12.3 Edinburgh Quay development as it appears from the historic Leamington Lift Bridge. Crowds are beginning to gather for the annual canal festival. Mounted on the construction fence at left, about half way down, are the wall panels recounting the neighborhood's industrial past. Ruderal vegetation peaks above the fence at far left (an indication of the length of time the site has been vacant).

Photo by Ingrid Buckley, 2014

Reflecting on the group's successes, one interviewee, who is a relative newcomer to the neighborhood, commented that it was, "A microcosm of how you'd want the wider urban realm to work." A long-time resident and FCI member stated that,

> The "anything goes" mentality (coming mainly from the economic development officials in the Council) is being tempered in Fountainbridge because of the high profile we are managing to sustain, the insistence we make for a model of place making, and quarterly 'Sounding Board' meetings between FCI, Council officials, and elected politicians.

The key is the level of involvement from residents:

> We did a lot to get the local people in. We drew people in from all sorts of organizations. That's what's really great about FCI and The Grove. They're really, really local. . . . We've had meetings with 80–90 people in them.

Residents may not have achieved perfection in Fountainbridge, but greater participation in the planning process is giving them a more hopeful vision for the future.

Patrick Geddes and conservative surgery

I recall many years ago perusing the pages of *Qualitative Methods in Human Geography*, edited by John Eyles and David M. Smith, and encountering for the first time a chapter provocatively titled "Topocide: The Annihilation of Place." Written by J. Douglas Porteous, it tells the story of Howdendyke, a small village in East Yorkshire, that was being systematically destroyed, slowly and painfully, by a "powerful politician-planner-industrial complex" (Porteous 1988, 91). Relying primarily on interview data, Porteous, who grew up in Howdendyke, describes how residents in this town of "72 inhabited houses and 11 acres of industry" responded to a 55-percent loss of housing and "three-fold increase in industrial space" between 1961 and 1985 (Porteous 1988, 80). His use of the term topocide, meaning "the deliberate annihilation of place," suggests more than the mere alteration of a physical landscape; it points to a situation where political and economic forces at the national scale cut residents out of the planning process, leaving them powerless to determine their own future (Porteous 1988, 75).

I was reminded of Porteous's chapter when I visited Fountainbridge for the first time. It was the summer of 2011 and I had just arrived in Edinburgh for a five-week study abroad trip. My family and I had rented a house in the Viewforth neighborhood just two blocks south of the Union Canal. From a third-floor window we watched as a wrecking crew cleared the remains of the McEwan's Brewery, a mainstay of the local economy for nearly a century and a half. As we wandered the streets we were disturbed by the scale and bleakness of the former industrial site (several city blocks flattened and fenced off). Our hosts, Jane and Lyn Jones, soon filled us in on the neighborhood's turbulent recent history. More importantly, they introduced us to the Fountainbridge Canalside Initiative. Unlike the villagers in Howdendyke 25 years earlier, residents of Fountainbridge were afforded an opportunity to assert themselves and their efforts are manifested in the landscape today.

Whether conscious of it or not, FCI and its allies appear to be inspired by the concept of "conservative surgery." According to Volker Welter (2002, 116), conservative surgery (as articulated by renown Scottish urbanist Patrick Geddes) means "amending and improving an urban quarter by minimizing the destruction of existing buildings, let alone the demolition of whole areas, for the sake of new houses and structures." It means preserving the built environment by adapting it to the needs of the present. And yet Geddes also recognized that historic vestiges might be replaced "without harming their ability to carry history onward," indicating he was more interested in the stories they tell than their authenticity (Welter 2002, 120). Although much of the built environment at its core has been razed, residents of Fountainbridge maintain an attachment to place and a dedication to the past that will likely ensure the survival (and perhaps even the revival) of this once thriving section of the city. They are living proof that history matters both inside and outside the boundaries of the World Heritage Site.

References

Abercrombie, P. and D. Plumstead. 1949. *A Civic Survey & Plan for the City & Royal Burgh of Edinburgh*. Edinburgh: Oliver and Boyd.

Alderman, D. and O. Dwyer. 2012. A Primer on the Geography of Memory: The Site and Situation of Commemorative Landscapes. In *Commemorative Landscapes of North Carolina*, edited by W. F. Brundage. (open access project) Chapel Hill: University of North Carolina at Chapel Hill. http://docsouth.unc.edu/commland/features/essays/alderman_two/

Alderman, D. H. and J. F. J. Inwood. 2013. Street Naming and the Politics of Belonging: Spatial Injustices in the Toponymic Commemoration of Martin Luther King, Jr. *Social & Cultural Geography* 14: 211–233.

Benton-Short, L. 2006. Politics, Public Space, and Memorials: The Brawl on the Mall. *Urban Geography* 27: 297–329.

Brown, B. B., D. Perkins, and G. Brown. 2003. Place Attachment in a Revitalizing Neighborhood: Individual and Block Levels of Analysis. *Journal of Environmental Psychology* 23: 259–271.

Daily Record. 18 June 2010. I Feel Lost, Admits Sean Connery as He Returns to Hometown Edinburgh. www.dailyrecord.co.uk/entertainment/celebrity/feel-lost-admits-sean-connery-1062103.

Dwyer, O. and D. Alderman. 2008. Memorial Landscapes: Analytic Questions and Metaphors. *GeoJournal* 73: 165–178.

Foote, K. E. 2016. On the Edge of Memory: Uneasy Legacies of Dissent, Terror, and Violence in the American Landscape. *Social Science Quarterly* 97: 115–122. DOI: 10.1111/ssqu.12259.

Foote, K. E. and S. Grider. 2010. Memorialization of US College and University Tragedies: Spaces of Mourning and Remembrance. In *Deathscapes: Spaces for Death, Dying, Mourning and Remembrance*, edited by A. Maddrell and J. D. Sidaway, 181–205. Farnham, UK: Ashgate.

Fountainbridge Canalside Initiative. 2012a. *Newsletter* March, 1–2.

Fountainbridge Canalside Initiative. 2012b. *Newsletter* May, 1.

Fountainbridge Canalside Initiative. 2012c. *Newsletter*. September, 1–3.

Fountainbridge Canalside Initiative. 2012d. *Newsletter*. November, 1–2.

Fountainbridge Canalside Initiative. 2013. *Newsletter*. April, 1.

Fountainbridge Canalside Initiative. 2014. *Newsletter*. February, 1–3.

Fountainbridge Canalside Initiative. 2015a. *Newsletter*. Winter, 1–2.

Fountainbridge Canalside Initiative. 2015b. Fountainbridge Community Wikihouse. Pamphlet.

The Grove Community Garden. 2014. Monthly Archives: February. https://grovecommunitygarden.wordpress.com/2014/02/.

The Guardian. 2016. Edinburgh's Age of Endarkenment: Development Is "Ripping Heart from City." 8 September. www.theguardian.com/cities/2016/sep/08/edinburgh-endarkenment-public-land-luxury-hotel-india-buildings.

Kallin, H. and T. Slater. 2014. Activating Territorial Stigma: Gentrifying Marginality on Edinburgh's Periphery. *Environment and Planning A* 46: 1351–1368.

Laxton, P. and R. Rodger. 2013. *Insanitary City: Henry Littlejohn and the Condition of Edinburgh*. Lancaster: Carnegie Publishing.

Macdonald, S. 2013. *Memorylands: Heritage and Identity in Europe Today*. London: Routledge.

Madgin, R. and R. Rodger. 2013. Inspiring Capital? Deconstructing Myths and Reconstructing Urban Environments, Edinburgh, 1860–2010. *Urban History* 40: 507–529.

Manzo, L. C. and D. D. Perkins. 2006. Finding Common Ground: The Importance of Place Attachment to Community Participation and Planning. *Journal of Planning Literature* 20: 33–50.

McKean, C. 2005. Twinning Cities: Modernisation versus Improvement. In *Edinburgh: The Making of a Capital City*, edited by B. Edwards and P. Jenkins, 42–63. Edinburgh: Edinburgh University Press.

Mesch, G. S. and O. Manor. 1998. Social Ties, Environmental Perception, and Local Attachment. *Environmental and Behavior* 30: 504–519.

Morris, R. J. 2007. The Capitalist, the Professor and the Soldier: The Re-Making of Edinburgh Castle, 1850–1900. *Planning Perspectives* 22: 55–78.

Morris, R. J. 2010. In Search of Twentieth-Century Edinburgh. *Book of the Old Edinburgh Club,* New Series Volume 8, 13–26.

The Old and New Towns of Edinburgh World Heritage Site Management Plan 2011–2016. https://www.ewht.org.uk/uploads/downloads/WHS_Management_Plan%202011.pdf

Porteous, J. D. 1988. Topocide: The Annihilation of Place. In *Qualitative Methods in Human Geography*, edited by J. Eyles and D. M. Smith, 75–93. Totowa, NJ: Barnes and Noble Books.

Post, C. W. 2016. Beyond Kent State? May 4 and Commemorating Violence in Public Space. *Geoforum* 76: 142–152.

Post, C. W. and D. H. Alderman. 2014. "Wiping New Berlin off the Map": Political Economy and The De-Germanization of the Toponymic Landscape in WWI USA. *Area* 46: 83–91.

Proshansky, H. M., A. K. Fabian, and R. Kaminoff. 1983. Place-Identity: Physical World Socialization of the Self. *Journal of Environmental Psychology* 3: 57–83.

Reeves, S. 2014. The Grove: Fountainbridge Community Garden. The Story So Far. (10 January). Pamphlet.

Rodger, R. 2005. Landscapes of Capital: Industry and the Built Environment in Edinburgh, 1750–1920. In *Edinburgh: The Making of a Capital City*, edited by B. Edwards and P. Jenkins, 85–102. Edinburgh: Edinburgh University Press.

The Scotsman. 2010. How the Site of the Fountainbridge Brewery Is Earmarked for an Ambitious Makeover. 3 August. www.scotsman.com/lifestyle/how-the-site-of-the-fountainbridge-brewery-is-earmarked-for-an-ambitious-makeover-1-819915.

Vaske, J. J. and K. C. Kobrin. 2001. Place Attachment and Environmentally Friendly Behavior. *Journal of Environmental Education* 32: 16–21.

Welter, V. M. 2002. *Biopolis: Patrick Geddes and the City of Life*. Cambridge: MIT Press.

13 Landscapes of recovery

Shifting senses of place attachment in Kesennuma, Japan

Rex "RJ" Rowley

On the afternoon of March 11, 2011, Japan experienced a magnitude 9.0 (Mw) earthquake. The resulting tsunami inundated coastlines, destroyed urban areas, and killed or displaced thousands of people in the Tohoku region. In the face of such disaster, outsiders may wonder, why would people want to continue living there? Or they may ignorantly assert, I would never move back to a place where something like that could happen again. Many resettled residents affected by natural disasters have posed similar questions (Samuels 2010). But, researchers report that residents who live in vulnerable places possess a different logic.

After the 2004 Indian Ocean earthquake and tsunami, thousands of survivors returned to rebuild cities and villages (Bird *et al.* 2007; Samuels 2010). Many observers expected the population of New Orleans to permanently decline following the diaspora of Hurricane Katrina in 2005, but reports of returning residents and rebounding neighborhoods are numerous (Chamlee-Wright and Storr 2009; Miller and Rivera 2010; Li *et al.* 2010). And, following a devastating tornado in Greensburg, Kansas, residents have not only rebuilt, but have made a conscious effort to make *how* they rebuild a reflection of their community (Paul and Che 2011; Smith and Cartlidge 2011).

Although many reasons exist for return, rebound, and reconstruction after natural disasters, an attachment to place is a central theme (Cox and Perry 2011; Silver and Grek-Martin 2015). Indeed, a sense of place, and the remaking of place plays a crucial role in disaster recovery (Entrikin 2007; Cox and Perry 2011). In early research following the 2011 Japanese earthquake disaster, sociologists, planners, and civil engineers made similar arguments for place attachment amid recovery (Hirano 2013; Miyake 2014). In one instance, residents saw their connection to place as a more important motivation for returning than economic forces (Ueda and Torigoe 2012).

An understanding of how a place recovers in the aftermath of a disaster can illuminate the indelible role that place plays in peoples' lives. By examining natural disasters, cultural geographers can broaden their research perspective beyond an examination of cultural meaning and social relations to also include the natural world (Entrikin 2007). In addition, amid disaster, the character of place "reinforces and deepens feelings of belonging and personal identity in the face of shared suffering" (Silver and Grek-Martin 2015, 33). Stated differently, recovery from natural disasters act as "an extreme, highly visible case in the otherwise unnoticed ebb and flow of the ongoing process of place making" (Entrikin 2007, 177).

The coastal port and fishing city of Kesennuma was one of the hardest-hit places during the Great East Japan Earthquake of 2011 (referred to simply as 3/11). The tsunami waves destroyed much of the city's commercial core and all of its low-lying coastal neighborhoods. Like New Orleans, Kesennuma faced the threat of permanent decline and loss. Notwithstanding overwhelming attention given to the disaster, research that highlights the role of place in post 3/11 Japan is rare (Karan and Suganuma 2016). I seek to add to the body of work that focuses on the slim but expanding geographies of disaster and place attachment by examining Kes-ennuma's landscapes of recovery. In particular, I explore two propositions from existing research: 1) that natural disasters, as extreme events, provide an important lens where the place-making process is made more visible and apparent (Entrikin 2007), and 2) that place identity and attachment are integral to and are, in fact, overtly emphasized by recovery efforts after a natural disaster (Chamlee-Wright and Storr 2009; Cox and Perry 2011).

The cultural landscape (built environment) is the material manifestation of local culture that relays messages of how residents and local leaders see their com-munity (Lewis 1979), or how they want outsiders to see their community (Mitch-ell 2008). Beyond its role as a reflection of culture, the cultural landscape, once formed, subsequently influences the people who interact with it (Schein 1997). In short, the cultural landscape is a visual library of human expressions that "embod-ies how people (through their experiences) connect these visible elements to place-based meanings and regional identity" (Wyckoff 2014, 4). Following a disaster, it is often the destruction, and eventual rebuilding of the cultural landscape that is the most noticeable indicator of impact. The built environment, then, is a vital clue to understanding the values and connections to place during recovery. Despite such usefulness, an analysis of the cultural landscape has served as a method of data collection in only a handful of post-disaster studies (see Miller and Rivera 2010; Samuels 2010; Smith and Cartlidge 2011).

In this chapter I use landscape analysis to reveal the character of place and the emotional ties residents have for place following a disaster. Data were gathered through participant observation and repeat photography of particular sites in Kes-ennuma during three visits over three years (August 2014 to May 2016). I supple-ment my findings with informal interviews with 12 residents who have stayed or returned. Through this methodological lens, scholars can identify elements of human interaction with place, how the characteristics of a place can foster strong emotional ties, and what meaning and value is drawn from human experience in that place (Tuan 1977). In what follows, I point to examples in the Kesennuma landscape that serve as symbols and sentiments of place attachment for local resi-dents amid recovery.

Presenting such evidence *during* the recovery (and place-remaking) process affords a unique glimpse into the important qualities of the place (Entrikin 2007). And, doing so over the course of three years while the community rebuilds has provided a unique "opportunity to watch the cultural landscape materialize at . . . a plotted pace" (Smith and Cartlidge 2011, 538). Indeed, I found the evidence about local attachment to place in Kesennuma hard to ignore. Still, such evidence must be considered amid "obvious shortcomings in any analysis of its interpretation since it [place] is still forming" (Entrikin 2007, 168).

My fieldwork focused on five study areas within Kesennuma (Figure 13.1). For simplicity, I have assigned names to each area based on their general location within the city. The Central Peninsula includes many coastal neighborhoods surrounding the main fishing and ferry ports, fish processing facilities, tourist, and other commercial zones. Two areas, Minami-Kesennuma and Matsuiwa, are named for neighborhoods near stations on the Kesennuma Rail Line. Hashikami and Ooshima are more straightforward geographic sections of Kesennuma, the former a rural area south of the main city characterized by a gradually sloping coastline, and the latter a rural island accessible only by boat. I refer to interviewees using pseudonyms to preserve their privacy. Following a short background on Kesennuma, I present my analysis of four categories of landscapes of recovery: *response*, *optimism*, *memory*, and *adaptation*. I then discuss some key lessons learned through my analysis.

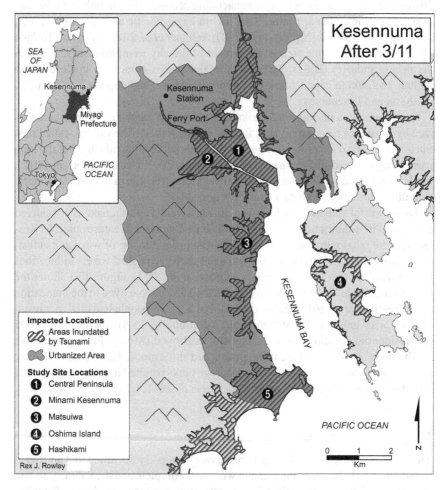

Figure 13.1 Map of study sites and areas in Kesennuma inundated by the 3/11 tsunami.
Sources: Japan Ministry of Land, Infrastructure, Transport, and Tourism (urbanized area); Japan Society for Geoinformatics (inundated areas).

Cartography by author

Background on Kesennuma

Kesennuma City is located in the far northeast of Miyagi Prefecture in Japan's Tohoku region (Figure 13.1). The municipality is comprised of numerous urbanized neighborhoods and rural villages that wrap around Kesennuma Bay. Mostly mountainous, the bulk of Kesennuma's population is scattered along a narrow coastal plain. Prior to 3/11, the 2010 Japanese Census reported a total population of 73,489 people. One year following the disaster the population had declined to 67,848, and by the 2015 census dropped to 64,917 (Godzik 2013; Japanese Statistics Bureau 2016).

The city's economy is reliant on fishing and tourism. The long, north-south-trending Kesennuma Bay hosts the largest landing harbor in Tohoku making it one of the nation's most important fishing hubs (Ueda and Torigoe 2012; Tomita 2014). The city is known for major aquaculture products that include tuna, bonito, oysters, scallops, clams, abalone, sea urchin eggs, Pacific saury, sea pineapple, shark fin, and seaweed (Nguyen *et al.* 2007; Yalciner *et al.* 2011; Ueda and Torigoe 2012). According to one source, 85 percent of the city's employment is based in the fishing industry (Biggs *et al.* 2011). Major tourist attractions in Kesennuma include Iwaisaki (a cape featuring a famous blowhole) and popular beaches (e.g., Ooshima and Oisehama).

Kesennuma lies on Japan's Sanriku coast, a rocky, jagged coastline characterized by dozens of penetrating bays formed in submerged river valleys, called rias. Such features provide protective harbor for a vibrant fishing industry, but the long narrow bays also amplify tsunami waves (Matanle 2011; Ueda and Torigoe 2012). Reclaimed land at sea level on the central peninsula (recognizable in the geometric shoreline) further exacerbate a tsunami's impacts.

The powerful tsunami that followed the massive 3/11 earthquake is considered a rare, one-in-a-thousand-year event. As such its damage was catastrophic. In Kesennuma the maximum inundation depth (from ground to top of wave at its crest) was measured at 10 meters, and the force of that wave pushed waters to locations as high as 23 meters above sea level (Suppasri *et al.* 2012). Figure 13.1 shows the areas inundated in Kesennuma. An estimated 40,331 people lived in the inundation zone (Yalciner *et al.* 2011) and as of April 2016, 1,214 people had died as a result of 3/11, with another 220 still missing. Additionally, 15,815 homes and 9,605 non-residential buildings were damaged or destroyed (Miyagi Prefecture 2016). Furthermore, nearly all of the fishing boats and ships moored at Kesennuma were lost, and fuel and oil spilled from their storage tanks resulted in a four-day fire that added to wave-wrought destruction (Yalciner *et al.* 2011; Ueda and Torigoe 2012).

Landscapes of response

The initial response to the disaster focused on cleanup. By the time I first arrived in the summer 2014, most of the cleanup work was complete and the fishing port had rebounded. Still, evidence of the initial destruction remained. Buildings that withstood the waves had yet to be removed, while others still showed the scars from wave-pushed debris scratching along their walls (Figure 13.2). Over the next two years, however, the recovery process I witnessed was astonishing. In 2015, new roads and sidewalks had been graded along with the foundations for new buildings. By 2016 many of those buildings were complete and the built environment became almost unrecognizable, particularly in the central peninsula.

This pattern of construction illustrates a common practice in post-disaster locations. The goal is to bring the city back to a (new) sense of normality (Miller and Rivera 2010; Samuels 2010; Cox and Perry 2011). Nowhere is this more evident than in Kesennuma's transportation landscape. The first and most visible aspects of the city to recover were the roads, and modifications to the public transportation system followed within a year. Prior to the disaster, a train line connected Kesennuma Station with neighboring cities. Due to the damage caused by the earthquake and tsunami, Japan Railways replaced this with a new Bus Rapid Transit (BRT), which follows the same general route as before. In rural Japanese communities like Kesennuma, access to public transportation is an integral part of life and mobility. The return of the Kesennuma Line BRT became a vital signal of this city's nascent reemergence.

The residential and commercial landscape has also begun to recover. Some residents have permanently left the city, and many simply moved to higher ground. Still others moved temporarily to neighboring communities, or to one of dozens of government-sponsored, prefabricated housing structures in local parks and schoolyards to wait for their new homes to be built. Such was the experience of

Figure 13.2 Photo of buildings on Kesennuma's central peninsula. The building to the right shows damage from debris contacting it during the 2011 tsunami. In the parcel adjacent to it is a *kasaage*.

Photo by author, 2016

the Yamano and Kamitani families who live in Minami-Kesennuma and Matsuiwa, respectively. Both families (and many others) waited months for the opportunity to return home. For them, returning to family-owned land that was also close to their place of employment was an important part of their personal recovery (see Ueda and Torigoe 2012).

Privately owned businesses have responded in a similar way. Soon after the massive cleanup, businesses began occupying temporary structures similar in design to the prefabricated houses. The restaurants, shops, and convenience stores I frequented adjacent to the Matsuiwa BRT station all were located in provisional buildings. One convenience store added a fake brick facade in an apparent attempt to provide a sense of permanence and normalcy for customers. At the ferry port in Sakanamachi, the *Kesennuma Yokocho Fukkou Yatai Mura* (a small complex of retail spaces located in temporary structures) was installed a few months after 3/11. The name of the shopping area translates to mean Kesennuma Streetside Recovery Stall Village, and it serves as a symbol of hope for local residents who now have local food markets and restaurants. Posted throughout the area are messages of encouragement for residents who are still coping with the disaster. One banner installed by Lions Club International displays the message, "Stay strong, Kesennuma. Stay strong, Tohoku. . . . The light of recovery will extend from these Yokocho stalls." Despite their temporary status, the presence of the Yokocho village not only signals the return of investment and income for entrepreneurs, but it also engenders hope that Kesennuma will emerge from the detritus left by the tsunami. Each landscape of response further underscores how Kesennuma residents care about their place and stand unwilling to simply let it vanish.

Landscapes of optimism

A sense of optimism permeates Kesennuma's response and recovery efforts. When I spoke to Mr. Kamitani in front of his newly rebuilt home, I asked if a repeat of the disaster concerned him. Noting his long-time family connection to this property, he responded with some concern, but said his property is a bit higher up in Matsuiwa neighborhood, and a 100-year tsunami won't affect him. He further explained that he knows he is not the only person in the world coping with a disaster. He noted Hurricane Katrina in New Orleans as one example. He was excited to show off his new two-story, western-style home. With yard decorations adorning the front of his house, his excitement at returning home was palpable. Mr. Wada, an older resident in Matsuiwa shared a similar story. He told me how he built a new home on his property within two years of 3/11. I got the sense that he was delighted to participate in the rebuilding of his neighborhood.

This is not to say that all is well. Clearly, Kesennuma is in the process of recovering from a terrible disaster, and some people feel a deeper level of concern than expressed by Kamitani or Wada. Such was the case with a group of older women I spoke to as they tended weeds on a once-occupied, but now vacant parcel in Hashikami. They shared a fear in returning to family homes. One woman explained the idea that young people have opportunities to leave and make a better life in larger cities, but that option is not available to older folks.

Despite an underlying concern, each time I returned to Kesennuma, I was struck by the positive energy permeating the community. A Zen Buddhist temple in Hashikami, for example, was partially destroyed by the tsunami as waves rolled over the area. Because the temple is located on higher ground, it was spared a total loss. Still, the water reached nearly nine feet above the main floor. Within a short period of time after the event, the temple had been completely restored. During my tour of the property, I noted a Japanese proverb on display in several locations. The temple master, Mr. Kawabata, explained that in the days following the disaster he wrote the words *Kujikenai, Nigenai, Megenai*, which translate to mean, "We will not shrink, we will not run away, we will not lose spirit." He added that we can't run away from nature; the sea gives us life even if it also takes on occasion. The motto of the temple is a powerful example and symbol of the hope that residents have in recovery.

One does not have to look far to find a host of other messages (from civic groups, private businesses, and local residents) announcing a similar attitude of fortitude, strength, and encouragement. A banner at the Kesennuma Yokocho market displays the following message: "2011.3.11. We will not forget the events of that day. We believe in a future of hope." I often saw bumper stickers proclaiming similar messages. One showed a caricatured shark with its humanized pectoral fin pumping the air and announcing (in Japanese): "Press Forward in the Recovery, Kesennuma!" Symbolizing a grassroots effort to promote optimism, another set of stickers encouraged a "prayer for recovery" accompanied by motivational colloquialisms, unique to this area (e.g., "Never give up! Kesennuma. Little by little, one step at a time, you'll make it.")

A number of uplifting messages include the image of Hoya Boya (Figure 13.3), a mascot created in 2008 for public relations and tourism development in Kesennuma (Kesennuma City 2016). Many cities and prefectures throughout Japan have developed such characters as a way to represent and market their community. *Hoya* is the Japanese word for sea pineapple (an edible sea squirt) and *Boya* is boy or child (e.g., the child of the sea). And, Hoya Boya's appearance literally represents the products local fishermen are known for: his head is a sea pineapple, his belt-buckle a scallop, his sword a Pacific saury fish, and he is often seen with a shark. The mascot has become more popular since the tsunami as local leaders have used him as a vehicle of encouragement as local residents work through the recovery. Hoya Boya's happy face is so pervasive in the community that it is difficult to turn anywhere without seeing him. He is on BRT busses, community signs, building sides, and safety fences at construction sites. His ubiquitous presence evokes a strong sense of place identity and optimism as Kesennuma recovers.

The Japanese word *ganbaru* is often used for encouragement for individuals facing some sort of challenge (e.g., work, sports, personal). It is ubiquitous on the signs and stickers displayed throughout Kesennuma. The word translates to "let's hold on" or "we can do it" or "go for it." Another, less common translation means "to remain in place," or "to stick to one's post." The full range of sentiments that ganbaru embodies is at the foundation of all words of encouragement expressed by residents of Kesennuma. It underscores a deep attachment to place that permeates the recovery process.

Figure 13.3 Photo of Hoya Boya alongside a line marking the height of the tsunami on
Kesennuma's central peninsula, near the ferry port. Such lines can also be
found on new buildings and inside reconstructed or renovated structures.

Photo by author, 2014

Landscapes of memory

Woven throughout landscapes of response and optimism are other elements that
evoke memories of the disaster. We often embed reminders of significant events
of the past in our built environment. Such landscapes, including those created
in response to a disaster, contribute to the place-making process. Landscapes of
memory can: help individuals in post-disaster places cope with tragedy (Entrikin
2007), remember the happy times before the event (Chamlee-Wright and Storr
2009), mourn and memorialize those lost (Samuels 2010; Nakamura 2012), and
plan for future threats (Ueda and Torigoe 2012).

The cultural landscape in Kesennuma elicits such emotions for residents. Nearly
everyone I spoke with was quick to share their stories of 3/11. Many lament the
loss of cherry trees along the estuary in Minimi-Kesennuma, of the destroyed
grave site of a relative, of an attractive tourist beach at Oisehama, and of a residential scene amid a backdrop of lapping waves at Hashikami. Four interviewees
spoke at considerable length of their personal story of escape and rebuilding. I will
share one representative example.

At the time of the quake and tsunami, Mr. Matsuyama was at Kesennuma Station, which sits at an elevation unaffected by the waves. But, he recalled, his wife
was in the lower part of the city. She got in her car and made it to the city hall

building where she could call her husband and notify him that she was safe. Still concerned for her life, she ran up another road and arrived at a hilltop school. As he shared the story, he gave me a set of hand-drawn maps showing the extent of the waves in his neighborhood (in Hashikami) and the path his wife took to escape. He pointed out that had she taken another route, she would have been stuck in traffic and would probably have died (all of these scenarios were carefully drawn and explained on his map).

What struck me most about Matsuyama's story were not the details expressed, but how much he *wanted* to share his memories of the day. Along with the maps, he had prepared a packet that included several news stories and a full-color publication of photos taken during and immediately after the disaster. He had prepared the packet for anyone interested in the events. Matsuyama was not alone; others were eager to share their personal photographs too. In his reconstructed home, Mr. Yamano prominently displays images of what the tsunami did to his previous residence. Likewise, temple master Mr. Kawabata guided me around one hall of the temple where photos of destruction and reconstruction hung next to precious tapestries and various posters. He also showed me a video documenting some of the news stories that focused on his particular part of the city.

The fact that so many people relived and freely shared their survival stories and photographic memories with me is indicative not only of the indelible impact of that day's events, but of their desire to remember this life-changing, place-changing moment. I was witnessing how "narration of threat and its association with place is part of the ongoing process of place making" (Entrikin 2007, 177).

Landscapes of memory are not restricted to private individuals. A number of physical remnants of the destruction still existed within Kesennuma. A ladle, a broken casing of a cassette tape, the shell of an electronic device, and pieces of roofing tiles lie among the weeds and rocks on a soon-to-be reutilized building foundation. Each stores memories of place. More striking are the formal monuments. I saw a number of such memorials in the small cemetery outside of a temple in Hashikami, including a grove of commemorative trees, a place set aside for prayers, a large stone with the names and ages of the victims, and a map showing the structures destroyed or flooded in the vicinity of the temple. Similar monuments have been erected at *Kotohira Jinja* (a Shinto shrine) near Iwaisaki and at a temple in Matsuiwa. In a vacant lot near the ferry port, fishing implements and driftwood were used to construct a temporary memorial that was labeled "Ground Zero" (in English). The makeshift memorial was removed after my 2015 visit to make room for a new construction project.

Perhaps the most prevalent, and prominent symbols of local memory are markers affixed to interior and exterior walls of buildings showing the level to which the water rose. Such markers are impossible to miss on a visit to the city's central peninsula. They typically show a line, the date of the disaster, a phrase saying something to the effect of "line of the Great Tsunami," and a measurement (e.g., 6.3 meters) for that particular spot. Often the lines are accompanied by the image of a Hoya Boya (Figure 13.3). Beyond building markers, on Highway 45, running along the coast, a prominent road sign shows the start and end of the line of tsunami

inundation. While serving as a reminder and warning for current and future generations, these markers also symbolize a process of place-making that acknowledges
the vulnerability that comes from living in a community so connected to the sea.

Landscapes of adaptation

Local residents and government officials are already planning for the next tsunami.
Existing scholarship finds that such adaptation is a common component in the
recovery process (Bird *et al.* 2007; Samuels 2010; Silver and Grek-Martin 2015).
Throughout Japan, the government has issued national guidelines and rebuilding
recommendations as countermeasures for future tsunami scenarios (Hirano 2013;
Tomita 2014). Evidence of such measures is ubiquitously found on the Kesennuma cultural landscape, most notably in new construction on higher ground.
Throughout the city, once forested hilltops have been cleared, making way for new
neighborhoods. In the city's low-lying areas, I saw new hotels and homes built
upon stilts. I also saw construction of dozens of *kasaage* (pedestal-like berms of
dirt and rock to raise the foundations for future buildings in the hardest-hit areas
of Kesennuma) (see Figure 13.2). This widespread adaptation strategy in Tohoku
(Hirano 2013) will bring a striking change to the look of the city.

The construction of a seawall is another recognizable adaptation effort. In 2014,
the Japanese government announced plans to fund and build this "Great Wall of
Japan" along the Tohoku coast. Despite deep concerns for their own safety, Kesennuma residents strongly criticized the project. One representative of the local
chamber of commerce noted, "If the seawall is too high, people (at the port) would
feel as if they were living in a prison. Many people would desert the town and it
could destroy the community" (Yoshida 2014). The seawall could also become
a nuisance for fishermen as they go about their work (Hirano 2013). One local
resident said: "We've never had a sea wall before. We've always coexisted with
the sea and we don't like being cut off from it" (Craft 2014). In a compromise
response, the Japanese government reduced the wall's planned height from 5.1 to
4.1 meters and raised the adjacent ground (with *kasaage*) by 2.8 meters (Yoshida
2014). The controversy surrounding the construction of the seawall underscores
the strong attachment local residents have for their seaside community. Protecting
Kesennuma is at the forefront of people's minds, but they don't want to give up
their identity in the process.

The construction of new high-rise apartments is another adaptation that does
not fit with the city's character. Although common in larger Japanese cities, these
large apartment buildings are changing the skyline of this fishing village. During
my fieldwork in 2014, two interviewees pointed out the new complex of three
high-rise apartment buildings under construction in their neighborhood in Minami-
Kesennuma. Now complete, the massive Nango Residence complex consists of
three main buildings (two six stories high and one ten stories high) and a central-
ized meeting area (Figure 13.4). No apartments exist on the first floor and tsunami
escape-route markers are found throughout the complex urging residents to take
refuge on top levels of the building in the event of a tsunami. In 2016, I saw five

Figure 13.4 Nango Recovery Housing near Minami-Kesennuma Station. These high-rise apartment complexes (common in larger Japanese cities) represent a new, post-tsunami housing architecture in Kesennuma. Note the lack of living space on the bottom floor and the small sign at the top of the left building indicating a tsunami escape area.

Photo by author, 2015

additional complexes of similar size complete or under construction in other parts of the city. Although a practical move on the part of city leaders given the low elevation of the city, the massive structures are out of place compared to the smaller homes and two-story apartments typical in this city. Such adaptation measures further highlight the fact that Kesennuma people desire to remain in their place following the disaster, even whilst accepting some change in the character of their city.

Place attachment amid recovery

Given the intense destruction following 3/11, with much of the cultural landscape and economy washed away, Kesennuma faced catastrophic loss. But, the people's desire to remake their place and *remain* has fueled a stunning recovery. An attachment to home, resource, and place-based identity has been a driving force in that effort, which is evident in the four landscapes of recovery I have highlighted here. A response to 3/11 has reshaped the built environment and restored some semblance of normalcy to the area, underscoring deep feelings of home that the people of Kesennuma possess. Optimistic messages on the landscape stem from local and regional leaders who display inspirational messages in public space and construct museum exhibits as well as from local residents who affix uplifting bumper stickers to their cars. Memorials and monuments stand as a testament to what has happened and the hope of what is yet to come, each existing in place as a permanent reminder of the city's sometimes tenuous connection to the sea. And, adaptation strategies have brought about a new and modified cultural landscape of hope for a present and future life in Kesennuma in line with both recovery and mitigation for future tsunami scenarios. Taken together, these landscapes suggest a reemphasized role of place in the recovery that has deepened feelings of attachment as some elements of the community have vanished and others have changed or been replaced. This aligns with scholarship noting how place attachment can become more evident and intense when a place is threatened or lost (Milligan 1998; Burley *et al.* 2007). I point to two examples illustrating this pattern in Kesennuma.

Kesennuma's strong connection to the sea has always been a part of its character. In the aftermath of the tsunami, the rebuilt cultural landscape speaks to a renewed and deepened sense of connection. A prime example is the character of Hoya Boya. Not only does he represent the economic and life-giving elements of the sea, but his visual appearance is also used as the face of the recovery. Likewise, ocean-themed posters or stickers, a prominent "Life with the Sea" exhibit at the reconstructed Shark Museum, and the comments of interviewees also attest to Kesennuma's connection to the sea. Mr. Katayama's remarks are representative of what I heard numerous time, "We must have respect for the ocean and the waves. The sea allows us to live." Other landscape features such as monuments, seawalls, raised building pedestals, and 3/11 waterline markers further reflect the city's relationship to the sea and the vulnerability that comes with it. That these features constitute a new and lasting everyday landscape for residents (in many instances springing from grassroots efforts of residents) speaks to an unambiguous symbol of the attachment that Kesennuma people have to their place.

A second, related example of this heightened and lasting sense of attachment to place is found in an examination of *fukkou* (a Japanese term that permeates much of the landscape and discourse surrounding the recovery). Typical meanings for this word include "recovery," "revival," or "reconstruction." But, directing attention to the two kanji characters that make up *fukkou* (復興) provides a more enlightened view of the term. The first of the two characters means "restore" or "revival," and the second, "interest (in something)" or "become prosperous." Hirano (2013) has compared this term to the "building back better" phrasing that has been used by the United Nations in other globally significant disasters (see Samuels 2010). It is no wonder, then, that the word *fukkou*, with its positive, optimistic connotation is seen and used throughout Kesennuma.

During interviews with local residents, I learned that the use of *fukkou* relates to the process and progress of recovery. Two markets housed in temporary structures near the main port used the word in their name (I have already discussed Yokocho Fukkou Stall Village, but another just two blocks from there is called *Kesennuma Fukkou Shopping Lane*). I met one shop owner in the latter location who sells souvenirs and a variety of Hoya Boya gear for tourists. He calls his store simply *Fukkou*, which he translates on store literature to mean: reborn. Several of the bumper stickers I discussed above employ the term in their optimistic messages. In several usages of the word, the two kanji were even changed to homophone characters to yield an identical reading but different meaning. One such instance was the bumper sticker with the triumphant shark encouraging perseverance. Here the second character in *fukkou* was substituted for one meaning "happiness" (復幸). In a similar move, two other temporary shop areas (away from the port) bore the names of *Tanaka Road Fukkou Village* and *Matsuiwa Fukkou Mart* using altered forms of *fukkou*. In both instances, the first *and* second characters were substituted with homophone kanji meaning "fortune" and "happiness" (福幸).

Application of the (unmodified) term occurred in more permanent instances. At the entrance to the new high-rise apartment complexes described earlier are signs describing these buildings as part of the *Kesennuma Fukkou Residences*. Similarly, a new sign at Takahama Beach on Ooshima describes this area as part of the *Sanriku Fukkou National Park*. The ubiquitous usage of this word seems to say, optimistically, "Yes, we are recovering from a great disaster, and we may never be the same. But we will push ahead, recognizing how far we've come, to a brighter future." Hirano sees *fukkou* as a part of the *machizukuri* (community development) trend in Japanese planning in recent years and noted, "the [meaning] of 'fukkou'. . . might be stronger place attachment and pride in the hometown even in post-tsunami reconstruction" (Hirano 2013, 5). The evidence of the disaster on the landscape not only serves as a reminder of what happened, but also of the connection residents have to this place and why they stay.

As the people of this city continue their efforts, attachment to and identity within place will continue to shift as a result of the disaster and recovery. In short, Kesennuma is a place that will never be the same again. This work has illustrated, through the lens of the cultural landscape, a shifting sense of place that can potentially serve as a symbol and a signal for other places that undergo drastic changes at the hands of disaster. Perhaps Kesennuma's landscapes of recovery may yield important answers for other places around the world facing a permanent alteration in a relationship to the sea. Geographers may, for example, explore the shifting place attachment of the peoples of low-lying atoll states in the Pacific who face the threat of sea level rise and climate change: how is a culture's place-based identity affected by the loss of homeland, displacement, and resettlement?

This portrayal of connectedness to place in Kesennuma also has broader implications beyond landscapes of recovery to the fundamental attachment each of us has to place. We all draw connections to our places, through lived experience, whether that be related to economics, emotions, or belonging. Perhaps an examination of landscapes of recovery and the extreme circumstances that bring place attachment to the fore can give us another glimpse into the power of the interaction between people and their places.

References

Biggs, S., K. Matsuyama, and F. Balfour. 2011. Tsunami Speeds "Terminal Decline" of Japan's Fish Industry. *Bloomberg News*, 25 April. www.bloomberg.com/news/articles/2011-04-24/tsunami-speeds-terminal-decline-of-japan-s-fishing-industry.

Bird, M., S. Cowie, A. Hawkes, B. Horton, C. Macgregor, J. E. Ong, A. T. S. Hwai, T. T. Sa, and Z. Yasin. 2007. Indian Ocean Tsunamis: Environmental and Socio-Economic Impacts in Langkawi, Malaysia. *The Geographical Journal* 173: 103–117.

Burley, D., P. Jenkins, S. Laska, and T. Davis. 2007. Place Attachment and Environmental Change in Coastal Louisiana. *Organization & Environment* 20: 347–366.

Chamlee-Wright, E. and V. H. Storr. 2009. "There's No Place Like New Orleans": Sense of Place and Community Recovery in the Ninth Ward after Hurricane Katrina. *Journal of Urban Affairs* 31: 615–634.

Cox, R. S. and K. E. Perry. 2011. Like a Fish Out of Water: Reconsidering Disaster Recovery and the Role of Place and Social Capital in Community Disaster Resilience. *American Journal of Community Psychology* 48: 395–411.

Craft, L. 2014. In Tsunami's Wake, Fierce Debate over Japan's "Great Wall". *NPR*, 11 March. www.npr.org/sections/parallels/2014/03/11/288691168/in-tsunamis-wake-fierce-debate-over-japans-great-wall.

Entrikin, J. N. 2007. Place Destruction and Cultural Trauma. In *Culture, Society and Democracy: The Interpretive Approach*, edited by I. Reed and J. C. Alexander, 163–179. New York: Routledge.

Godzik, M. 2013. Rebuilding Housing in Japan's Tsunami-Hit Towns and Cities. *International Journal of Housing Policy* 13: 433–445.

Hirano, K. 2013. Difficulties in Post-Tsunami Reconstruction Plan Following Japan's 3.11 Mega Disaster: Dilemma between Protection and Sustainability. *Journal of JSCE* 1: 1–11.

Japan Statistics Bureau. 2016. Population Census. www.stat.go.jp/english/data/kokusei/index.htm.

Karan, P. P. and U. Suganuma. 2016. *Japan after 3/11: Global Perspectives on the Earthquake, Tsunami, and Fukushima Meltdown*. Lexington: University of Kentucky Press.

Kesennuma City. 2016. Child of the Sea, Hoya Boya. www.city.kesennuma.lg.jp/www/contents/1232593384266/.

Lewis, P. F. 1979. Axioms for Reading the Landscape: Some Guides to the American Scene. In *The Interpretation of Ordinary Landscapes*, edited by D. W. Meinig, 11–32. New York: Oxford University Press.

Li, W., C. A. Airriess, A. C. Chen, K. J. Leong, and V. Keith. 2010. Katrina and Migration: Evacuation and Return by African Americans and Vietnamese Americans in an Eastern New Orleans Suburb. *Professional Geographer* 62: 103–118.

Matanle, P. 2011. The Great East Japan Earthquake, Tsunami, and Nuclear Meltdown: Towards the (Re)construction of a Safe, Sustainable, and Compassionate Society in Japan's Shrinking Regions. *Local Environment* 16: 823–847.

Miller, D. S. and J. D. Rivera. 2010. Landscapes of Disaster and Place Orientation in the Aftermath of Hurricane Katrina. In *The Sociology of Katrina: Perspectives on a Modern Catastrophe*, edited by D. L. Brunsma, D. Overfelt, and J. S. Picou, 177–189. Lanham, MD: Rowman & Littlefield Publishers, Inc.

Milligan, M. J. 1998. Interactional Past and Potential: The Social Construction of Place Attachment. *Symbolic Interaction* 21: 1–33.

Mitchell, D. 2008. New Axioms for Reading the Landscape: Paying Attention to Political Economy and Social Justice. In *Political Economies of Landscape Change*, edited by J. L. Wescoat and D. M. Johnston, 29–50. Dordrecht: Springer.

Miyagi Prefecture. 2016. Earthquake Damage Information. www.pref.miyagi.jp/kinkyu.htm.

Miyake, S. 2014. Post-Disaster Reconstruction in Iwate and New Planning Challenges for Japan. *Planning Theory & Practice* 15: 246–250.

Nakamura, F. 2012. Memory in the Debris: The 3/11 Great Japan Earthquake and Tsunami. *Anthropology Today* 28: 20–23.

Nguyen, T., N. Taniguchi, M. Nakajima, U. Na-Nakorn, N. Sukumasavin, and K. Yamamoto. 2007. Aquaculture of Sea-Pineapple, Halocynthia Roretzi in Japan. *Aquaculture Asia* 12: 21–23.

Paul, B. K. and D. Che. 2011. Opportunities and Challenges in Rebuilding Tornado-Impacted Greensburg, Kansas as "Stronger, Better, and Greener". *GeoJournal* 76: 93–108.

Samuels, A. 2010. Remaking Neighbourhoods in Banda Aceh: Post-Tsunami Reconstruction of Everyday Life. In *Post-Disaster Reconstruction: Lessons from Aceh*, edited by M. Clarke, I. Fanany, and S. Kenny, 210–223. New York: Earthscan.

Schein, R. H. 1997. The Place of Landscape: A Conceptual Framework for Interpreting an American Scene. *Annals of the Association of American Geographers* 87: 660–680.

Silver, A. and J. Grek-Martin. 2015. "Now We Understand What Community Really Means": Reconceptualizing the Role of Sense of Place in the Disaster Recovery Process. *Journal of Environmental Psychology* 42: 32–41.

Smith, J. S. and M. R. Cartlidge. 2011. Place Attachment among Retirees in Greensburg, Kansas. *Geographical Review* 101: 536–555.

Suppasri, A., S. Koshimura, K. Imai, E. Mas, H. Gokon, A. Muhari, and F. Imamura. 2012. Damage Characteristic and Field Survey of the 2011 Great East Japan Tsunami in Miyagi Prefecture. *Coastal Engineering Journal* 54: 125–155.

Tomita, H. 2014. Reconstruction of Tsunami-Devastated Fishing Villages in the Tohoku Region of Japan and the Challenges for Planning. *Planning Theory & Practice* 15: 242–246.

Tuan, Y.-F. 1977. *Space and Place: The Perspective of Experience*. Minneapolis: University of Minnesota Press.

Ueda, K. and H. Torigoe. 2012. Why Do Victims of the Tsunami Return to the Coast? *International Journal of Japanese Sociology* 21: 21–29.

Wyckoff, W. 2014. *How to Read the American West: A Field Guide*. Seattle: University of Washington Press.

Yalciner, A. C., C. Ozer, A. Zaytsev, A. Suppasri, E. Mas, N. Kalligeris, O. Necmioglu, F. Imamura, N. M. Ozei, and C. Synolakis. 2011. Field Survey on the Coastal Impacts of 11 March 2011 Great East Japan Tsunami: Seismic Protection and Cultural Heritage.

Yoshida, R. 2014. Tohoku Finding Real Recovery Hard to Come By. *The Japan Times*, 5 March. www.japantimes.co.jp/news/2014/03/05/national/for-kesennuma-real-recovery-may-never-come/.

Epilogue
Methodologies of place attachment research

Paul C. Adams

The previous chapters in this book have taken us many places, dropping in on fascinating manifestations of place attachment: from big cities to small towns to wilderness areas, from an island in the South Pacific to the French countryside to the mountains of the American West. We have visited places beset by violence, poverty, industrial decline, and natural hazards, but also places associated with the thrill of athletic competition, cherished memories, protection from the world, and a sense of connection to the planet and cosmos. We have seen how a sense of place can support a sense of time: a lost golden era, a new life, a vanishing tradition, a shadowy past, a marketable legend, a rebirth from the rubble. With these place encounters in mind, this epilogue returns to the important questions Jeffrey Smith raised at the outset regarding the study of place attachment, and reflects on the research methodologies that have been used get a handle on this tricky issue of place attachment.

"Place attachment" can sound a bit clinical and it helps to recall Yi-Fu Tuan's felicitous term, topophilia. There is something provocative yet compelling about the idea that what we feel for a place rises to the level of love rather than mere dependency or affinity. Tuan asserted that topophilia is "diffuse as concept [yet] vivid and concrete as personal experience" (1974, 4). Of course, personal experience can be atopic, that is, place-denying, inducing us to move to a bigger house every few years, vacation in kitschy tourist-oriented resorts, and envision ourselves as cyborgs dwelling in virtual realms of cyberspace. However, the research in this volume reminds us of the pull exerted by real, unique, physical places of all scales and types, including hometowns and neighborhoods, workplaces (including abandoned ones), sports stadiums, hiking trails, rivers, and memorials, to name only a few examples.

As Smith indicates, the topophilic pull can be inward, toward nurturance and security, or outward, toward novelty. He suggests as well that place is loved for its powers of restoration, validation, and transformation. But love of any kind places its bearer at risk of heartbreak. Places change! And whether this happens through the cataclysmic energies of nature or the gradual and sporadic impacts of economic restructuring, topophilic suffering arises when a place we knew and loved no longer exists as it was. So Smith rightly includes "vanishing places" among the basic experiences of place attachment, just as the potential of loss is always present in our attachments to other humans.

Building on Smith's arguments, the agenda for place attachment research includes versions of the classic questions addressed by journalists. *What* exactly is place attachment? Is it a feeling? An attitude? A way of being? A mode of interacting with one's environment? *Who* can we turn to for insights into place attachment? Who makes up the human groups we should attend to as relevant for place attachment research? *Why* should the effort be made to subject something as elusive as place attachment to academic scrutiny? Is our purpose to analyze? To empathize? To explain? To facilitate? To intervene? *How* should we go about doing topophilia research? What specifically should be our methodologies? How can scholars come to grips with the subtle and elusive experiences that fall under the rubric of topophilia?

This epilogue focuses primarily on the *how* questions in order to gain insight into the *what, who,* and *why* questions. Most of the chapter delves into methodologies of place attachment research that have been demonstrated throughout the book as they provide jumping off points to the larger implications of place attachment research. To begin unpacking the complexity of place attachment, however, we first consider the existential continuum from insideness to outsideness and how the study of place attachment presents the specific challenge of moving systematically from an outside perspective to an inside perspective. In the conclusion I advance a more integrative reflection on methodology, returning to the *what, who,* and *why* questions.

Insideness, outsideness, and researcher positionality

In Relph's memorable typology (1976), existential insideness marks one extreme of place experience. It is the place experience of someone saturated by a place through a long period of involvement. The opposite extreme is existential outsideness. In connection with the first of these, we can think of a small town – a traditional community where everyone knows everyone else, where many of the residents are second or third cousins, where nearly everyone goes to the same church, where driving directions assume you know which way is north, south, east, and west, where everyone calls a road "Dutch Avenue" or "Main" even though on the road signs it is marked "E. 82nd Ave" or "Alternate US Highway 90." The overwhelming majority of the town's residents are existential insiders, so deeply embedded in the place that it is only with difficulty that they can share their place knowledge with outsiders. While such places are becoming rarer, they are where it is easy to find existential insiders.

At the opposite extreme is existential outsideness – a condition in which one engages with a place in a thoroughly alienated way. I think of a visit I paid to an acquaintance in Brooklyn when I was just out of college. Brooklyn felt surreal, gray and inhuman, full of imagined and real threats from parking tickets to muggings. I recall waking up before anyone else in the apartment, stepping out onto a narrow concrete balcony facing out across a dirty jumble of buildings, leaning against a rusted railing and listening to the blaring horns and sirens from the streets below. At that moment I felt that life was an exceedingly grim business. Many people have had similar moments when they long for something welcoming or familiar, and when they are, in the words of Bob Dylan: "like a complete unknown, like a rolling stone."

A researcher studying place attachment must start as an outsider, exposed at times to jarring feelings of existential outsideness. Successful research crosses an immense gap to reach some sort of insideness (though rarely existential insideness) and then convey it to others. This crossing must occur in a relatively short amount of time, as these things normally go, and it usually raises more questions than it answers. What is required to sense what a place means to insiders? How can one encounter the place on its own terms? Place attachment research requires access to the insights of those who have lived in the place for a long time but these potential respondents have absorbed the place so deeply they cannot easily express their insights. Place research involves starting as an outsider and groping one's way through the dark toward the sense of place and aspects of locale (social place) that an insider cannot easily describe (Adams 2017). Place scholars therefore differ from ordinary inhabitants of a place in that they employ self-conscious methods to become aware of a place in a way that goes beyond their own personal relationship to the place. Their relationship to place is a hasty but deliberate journey away from existential outsideness toward an insideness that can be analyzed, interpreted and shared. If one is "just passing through," one cannot learn a huge amount about a place; yet here one is, a stranger in a strange land with a deadline and a notebook.

Drawing again on Relph (1976), there are at least three distinct pathways for self-conscious entry into place experience. Each of these aligns with certain research methodologies. First, there is *vicarious insideness*: the sense of place one can access through representations such as novels, poetry, photographs, and films. One sees the place through the eyes of another, hears through their ears, and benefits above all from their creative talents. Anyone can experience place in this way by reading a good novel or watching a movie, but scholars employ special tools to dig into place representations (Tomaney 2012). Content analysis and discourse analysis are included in the toolkit while the archive is one of its major "study sites." The associated techniques reveal discourses of all kinds as carrying relevant place representations which can be identified through techniques ranging from critical theory to semiotics to hermeneutics.

Second, there is *behavioral insideness*. This is a matter of having the know-how to navigate the roads and footpaths, find things, and generally orient oneself in a place. This form of insideness emerges as one inhabits a place. It begins after a few days in the field and it can be more fully developed after a few weeks in the field, although it takes years to learn all of the ins and outs of a local landscape. Landscape analysis is one methodology that offers behavioral insideness, reading visible features such as roads and buildings as signs and traces of the routines of inhabitants and their habits of dwelling in that place. Participant observation is another powerful methodology that builds behavioral insideness while cultivating other forms of place-appreciation as well.

Third, there is *empathetic insideness*, understanding of place which forms part of a more general understanding of its inhabitants, how they feel about the place, and what daily life feels like to an inhabitant of the place. A place attachment researcher can glimpse this aspect of place by cultivating understanding that goes beyond the sum of people and place and develops an understanding of *people in place*. Popular methodologies which develop empathetic insideness are the survey, the interview, participant observation and the focus group, all of which tap into the peculiar mix of feelings or feeling tone that binds insiders to the place. The utility of a survey depends on whether the researcher can read between the lines of closed-ended and short answer responses. Interviews and focus groups expose more complex attitudes through open-ended reflection, conversation and interaction. Participant observation can carry a researcher even farther in this direction but it requires one to live in a place as the locals live in the place, experiencing the many aspects of daily life that support place attachment.

In condensed form, place-attachment research benefits from layers of understanding and involves: interpreting representations of the place (vicarious insideness), acquiring familiarity with dwelling and getting around in the place (behavioral insideness), and developing empathy for how the world looks and feels to inhabitants of the place (empathetic insideness). Each of the above methods promotes a different kind of insideness. In combination these approaches are far more useful than any one of them in isolation. A researcher who has developed all three aspects of insideness can figure out the best questions to ask of and about the existential insiders in a place in order to explore their kinds of topophilia. Recognition of particularity is what structures and binds studies of place in lieu of abstraction and generalization, standing in complementary relation to what Donna Haraway called the "god trick" of objectivity (Haraway 1988, 581).

Place attachment research must integrate both subjective and objective phenomena, including for example perceptions and actions, beliefs and behaviors, attitudes toward nature and impacts on nature. Place attachment research combines forms of abstraction with attention to particular phenomena, because topophilia itself occupies an ontological middle ground between the specific and the general, the concrete and the abstract, the objective and the subjective. A balanced approach therefore combines description and analysis, attending to the *ontological betweenness* of place itself (Adams 2017; Entrikin 1991; Sack 1980). But how might various research methods contribute to this endeavor in a particular site and situation? We turn now to look back at the chapters in this volume for insight into that question.

Geographical methodologies for place attachment research

Archival research

"The archive" used to be accessed by pushing through creaking doors into a shadowy realm permeated by the smell of old books (at least insofar as attention was generally directed at public rather than private archives, Delyser 2014a). Now "archival" data is accessed most often by typing a password on a computer. This brings to the fore questions of interpretation and generates some degree of blending between archival methodologies and discourse analysis (see as follows).

These changes also help us to view the archive as a living memory, collectively maintained and developed, available to embodied and mediated modes of engagement, yet also capable of supporting creative links between the present and the past. Just as in the past, however, the archive is an avenue for the researcher to develop a personal bond to the place he or she is studying (Delyser 2014b). As such, the archive is not just a source of information *about* topophilia; it is also a vehicle *for* topophilia.

To some degree, all of the contributors to this volume employ archival research insofar as that facet of research lays the groundwork for other research trajectories, pointing out the assumptions that can be made about a place and the questions that should be asked about it. Several of the chapters rely intensively on archival material and I have picked three to provide a closer look at archival research: Tyra Olstad's investigation of the management of the Adirondacks with attention to contested images of wilderness, Douglas Hurt's chronicle of the succession of St. Louis baseball stadiums with attention to the changing American city, and Engrid Barnett's examination of the tensions and reconciliations between the ideal and the real in Virginia City, Nevada. Throughout these works "the archive" offers valuable insights into the contradictions and complexities of place attachment.

Olstad's chapter reveals how powerful a single document can be. For thirty years the Adirondack Park State Land Master Plan (APSLMP) provided guidelines shaping management of the Adirondack region so as to "preserve, enhance and restore where necessary, its natural conditions" while promoting "primitive" and "unconfined" recreation. Administrative practices and material landscapes of the Adirondacks revealed inherent contradictions between the ideal of "unconfined" recreation and the objective of controlling human impacts. The archive's role in this study is to provide both the master plan and examples of the resulting management strategies as they relate to trails, campsites, and built structures. Participant observation supplements the archive as a way to reveal the inherent tensions within environmental stewardship as a form of place attachment.

In Hurt's chapter, the challenge is to demonstrate place attachment across three historical periods and three successive incarnations of the St. Louis baseball stadium. Hurt's study tracks a moving target: a "Busch Stadium" that is recreated three times, in different locations, with different physical forms, design philosophies, and financial arrangements. From many archival sources a narrative emerges that goes beyond the stories of Busch Stadium I, II, and III, revealing urban evolution through the sports stadium. This study suggests the need for historical geographies using the archive to delve into other kinds of relocated, rebuilt, and reincarnated places: relocated business offices, airports, freeways, recreation centers, concert halls, schools, libraries, and the like.

Barnett's chapter uses archival research (as well as interviews and landscape analysis) to encapsulate the historical geography of the former silver mining town of Virginia City, Nevada. Over time a succession of outsiders imposed on the original mining town their images of a mythical Comstock they hoped to find. The special challenge here was for the researcher to capture an inside perspective of a place that has developed under the influence of so many outsiders. Drawing inspiration from the discipline that is most at home in the archive, namely history, this chapter addresses tricky conceptual issues through a narrative that blends historiography and cultural geography in richly descriptive writing.

These three works point to some general principles for archival research: first, never underestimate the power of a single text or document to shape place attachment; second, recognize any place attachment as something that may have been transplanted from one location to another; third, consider the examples set by other disciplines that depend on the archive, insofar as a fine-tuned narrative can help construct a captivating story and deepen understanding. Most studies of place treat the archival component as a foundation on which to develop primary research, so we turn now to the use of primary data.

Landscape analysis

In 1979, Don Meinig demonstrated ten different ways of observing the same scene, variously viewing landscape as: nature, habitat, artifact, system, problem, wealth, place, ideology, history, and aesthetic. In evoking these different perspectives, Meinig argued that: "For those of us who are convinced that landscapes mirror and landscapes matter, that they tell us much about the values we hold and at the same time affect the quality of the lives we lead, there is ever the need for wider conversations about ideas and impressions and concerns relating to the landscapes we share" (Meinig 1979, 47). This classic work suggests that the attachments people feel to landscapes are reflective of the ways they see landscapes, and these perspectives emerge in turn out of particular ways of relating actively and materially to landscapes. The chapters by Jeffrey Smith, RJ Rowley and Yolonda Youngs each link place to landscape processes in this way. They tell us, respectively, of the collectively built and maintained facilities that make a Paraguayan slum livable; of a Japanese town where the post-tsunami reconstruction efforts call on patience, fortitude and a cartoon mascot; and of the fusion of a designated wilderness landscape with waterborne recreational mobility. All three of these studies take agency seriously, revealing how place attachment depends on people actively engaging with landscapes in ways that rework the meanings of both place and self. Let us look at each a bit more closely.

Smith's landscape study takes us into the informal settlement of the Chacarita, a Paraguayan neighborhood which is normally off limits to outsiders. Prevented by local residents and police from entering the community on his own, he makes the most of guided tours and even uses them as a means of acquiring semi-structured interviews. He has limited time in the study site and limited opportunity to explore, meet people, make discoveries, ask questions, follow up on leads, and really feel what this place is like as a home. Facing this challenge offers lessons:

> On numerous occasions I attempted to enter the Chacarita as I walked down a street or alley. Each time I was turned away by residents who informed me that outsiders are unwelcome from entering the Chacarita without being accompanied by a permanent resident.

This says a lot about what Jane Jacobs calls "eyes on the street" (1961), informal policing, which in turn attests to strong collective place attachment – defensive topophilia. Thus, place attachment research is informed not simply by site visits but also by the peculiar difficulties accessing a landscape, and how each visit is constrained or enabled.

Rowley captures three years of the recuperation and restoration in the wake of the 2011 tsunami that struck the northeast coast of Honshu, Japan. Here landscape interpretation in the city of Kesennuma discovers that what is being rebuilt is not merely physical infrastructure of the city but also the residents' sense of hope and community identity. A longitudinal methodology involving interviews, participant observation, and repeat photography reveals post-disaster changes not just in the material landscape but also in landscape-related feelings, symbols, and activities. There are many new buildings in Kesennuma but just as important is the appearance in the landscape of Hoya Boya, a cartoon sea pineapple with blobby orange siphons sticking out of his head. He smiles from walls, fences and vehicles, promoting affective responses of enthusiasm and optimism during the long and tedious rebuilding process. Rowley's analysis reveals how the physical landscape and attached symbols can work together.

In Youngs's chapter, landscape analysis is again combined with participant observation, as well as with archival research. Youngs focuses on river runners who offer tours on rubber pontoon boats in Grand Teton National Park. She demonstrates that place attachment can be created through mobility, as people engage with this rugged environment by floating down its turbulent rivers. Complementary perspectives could examine the topophilic sentiments generated by hiking, backpacking, mountain climbing, cross-country skiing, snowshoeing, mountain biking, snowmobiling, boating, and canoeing. Participant observation helps bring to this kind of research an understanding of the interplay between the geomorphic processes of a dynamic natural environment and the social processes of recreation, tourism, and nature preservation.

Throughout these various contributions we see how landscape interpretation has taken up the methodological challenge to go beyond a narrow focus on *landscape as form*. Landscape research is only partly about the shape of natural features or human creations like buildings, memorials, and cities. It also attends to embodied human actions occurring in the past, present, and future, engaging with *landscape as process*. The best landscape research works to integrate place and time, beyond mere vision and even beyond multi-sensory engagement, to propel the viewing subject toward what Relph calls vicarious insideness (1976). Summarizing: Smith indicates that a focus on landscape can bring aspects of place attachment into the open, including territoriality and policing; Rowley takes up the challenge of seeing how place attachment can guide landscape transformations that keep a place from becoming abandoned and disused; Youngs helps us see how place attachment is not necessarily a matter of being fixed to a particular location, but can also depend on moving through the landscape in a particular way. Building on these arguments, one's landscape affects who one is, establishing one's *position* in both literal and figurative ways. How can we get to know these subject positions better? Surveys, interviews, and focus groups provide a partial solution.

Surveys, interviews, and focus groups

An interview allows research subjects to express in their own words what a particular place means to them. Depending on the research question, it may in fact be quite appropriate to depend entirely on interviews as a source of data, for example when one is studying structures and dynamics of social processes and the meanings ascribed to them by people living in a place. Surveys can provide a more quantifiable, statistically comparable version of the same information, but it has been argued that whereas interviews are often buttressed by more "objective" methodologies such as surveys this habit may in fact be a "quantitative prop to provide an illusion of academic responsibility" (Winchester 1999, 66). In a recent progress report on qualitative methods, Dowling *et al.* (2015, 680) argued that: "Qualitative interviews – semi-structured or unstructured, with individuals or with groups – continue to predominate in the social and cultural geography subdisciplines." Barnett captures a common situation for human geographers working in the field, explaining that the "interviewing process comprised, by far, the largest chunk of research that I undertook and delivered the greatest dividends." In any case, words and worlds are complexly intertwined (Pred 1990).

Interviews enrich many of the studies in this collection. Smith hears from a resident of the Barrio that "Everyone who lives on the outside thinks this place is terrible. I love it here. It's close to the downtown and I know all of my neighbors." Ducros finds a French villager who argues that if people alter the appearance of their houses "we don't have a village anymore." One of the Pitcairners interviewed by Johnson worries that with children educated off the island, "we will forget the reasons behind the place names and the places will be lost forever." A resident interviewed by Rowley in post-tsunami Kesennuma says "We must have respect for the ocean and the waves. The sea allows us to live." Such quotes bring to life the multiple axes on which place attachment varies from place to place, person to person, and situation to situation.

Most research subjects need little encouragement to talk about the places they call home, and will speak with equal alacrity about places they have owned, built, renovated, navigated, defended, or maintained. Place attachment is a performative process insofar as one projects oneself out into a place to perform one's attachment, and speaking about place is yet another performance that both transforms and validates parts of oneself. As such, an interview can be less an imposition on the research subject than a welcomed event. While an interview is unmatched for depth, the question of breadth lingers. How much can we make of a few words about a place? Can we generalize from the quotes we obtain in interviews? There is no easy answer to these questions, so it is important to be judicious in regard to the selection and contextualization of quotes that will be shared in published work. In some cases the voice of the author will need to convey the story over and above the quotes; in others the voices of research subjects will come through more strongly than the author (or seem to, at least). A closer consideration of the chapters by Geoffrey Buckley, Hélène Ducros, and Michael Strong allows these ideas to be fleshed out a bit more.

Buckley allows the story of landscape change in Edinburgh's Fountainbridge neighborhood to be told through the words of local inhabitants. Whether conveying a memory of the pungent smell from the now-defunct brewery, or feelings of abandoned spaces – "sad, bleak, impersonal, scary, dead" – or extolling the social environment of a new garden with its "microcosm of how you'd want the wider urban realm to work," his interviews offer many glimpses of place change in this working-class neighborhood. His chapter shows that interviews can capture the good and the bad aspects of place in very particular terms, not merely situating the place in general processes like globalization and deindustrialization, but breathing life into a portrait of a place as the lived-in world of various inhabitants.

Ducros's study begins with participant observation, followed by interviews of staff in a grassroots organization, further interviews of the mayors of villages labeled as "Most Beautiful Villages," and a final round of interviews of over a hundred "key actors and residents" in such "labelized" villages. This systematic incorporation of perspectives from different kinds of stakeholders allows for a level of control since a given subject's interests and knowledge inevitably create an idiosyncratic contribution to the understanding of a particular place, and place attachment in general. The insights gleaned in one phase of the interview study with one group of subjects can be carried forward to the next phase in the shape of more and better questions. A researcher can respond more adroitly to the comments of later interviewees having assimilated the categories, assumptions, terms, constructs, attitudes, beliefs, frames, and biases of earlier respondents.

Strong demonstrates an approach to interview analysis that emphasizes short snippets and even individual words from various participants. These are juxtaposed in rapid succession, highlighting commonalities between subjects rather than unique articulations. Thus, we learn that "Many residents believed the resettlement was 'bad for everyone in the village.' Words commonly used to describe life in Bairro Chipanga include 'challenging' and 'insufferable'." In situations where translation is difficult it may be helpful to emphasize key terms and elucidate their meanings in the local context. When subject responses are taped, a long verbatim response is easier to capture than when a researcher is limited, for whatever reason, to taking notes. Such variables in the interview situation will therefore lead to rather different presentations of verbal responses and different types of quoted material in the resulting research.

An important variation on the interview is the focus group. The benefits of focus group research lend themselves well to the study of place attachment. Interaction among participants elicits ideas and emotions that would be difficult or impossible for the researcher to capture through one-on-one interviews or surveys. Contradictory place images are juxtaposed with each other in the contributions of various participants with different points of view. For example, Randy Peppler, Kimberly Klockow, and Richard Smith employed focus groups in their study of people's perceptions of tornado risk in central Oklahoma and included weather experts and non-experts. Place attachment is important to risk management insofar as people often disregard warnings regarding floods, wildfires, and extreme weather events because they feel most secure at home, even when their homes are directly at risk. The authors integrated experts such as meteorologists, emergency responders, and reporters into the focus group discussions, permitting non-experts to engage with these experts in a semi-guided discovery process. This methodology extended the focus group beyond data gathering to serve as an information channel to key local decision makers. People tasked with informing the public about risk were involved in the dialogic exploration of risk with other stakeholders, thereby closing the loop between research subject and research audience.

Discourse analysis

Geographers began to analyze literature, arts, and the humanities in the late 1970s and early 1980s, with inspiration from the philosophical traditions of hermeneutics and phenomenology (Tuan 1978; Pocock 1981). A more critical approach followed a decade later drawing on a wide range of theoretical infusions including the Frankfurt School, British cultural studies, and French social theory (Cosgrove and Daniels 1988; Barnes and Duncan 1992; Duncan and Ley 1993; Crang and Thrift 2000; Adams *et al.* 2001). Essentially any kind of cultural product, text, meaningful exchange, or performance is fair game for discourse analysis if it is explored through "processual engagement with a text" to reveal its "iterative, emergent, and dialogic properties" (Steacy *et al.* 2016, 166). Government publications can provide insight into official place images, while at the opposite extreme private communications such as journal entries can provide insight into some of the most personal aspects of place attachment. In the middle there is much research addressing films and television shows, online debates, magazines, books and other "mass" or popular media as clues to the ongoing evolution of discourse as a collaborative and contested endeavor. The degree of popularity of such material indicates if it has struck a chord with the public, and audience reception can be assessed in more specific ways to see what people think of such representations. To take discourse analysis seriously requires that we treat people's responses to mediated place representations as contributions to public discourse in their own right. Our own questions to subjects and the published works we develop from their responses are further discursive contributions. What is suggested here is that discourse analysis prompts a profound (re)orientation of understanding such that meaning is not something we point to with our scholarly communications but rather a process that circulates through all communications, including scholarly communications. Various contributions to this volume demonstrate how elements of place attachment can be discerned through analysis and critique of various texts and discourses. We will focus on the contributions of Steven Schnell, Christine Johnson, and Chris Post.

Schnell offers a close reading of the graphic novel, *The Arrival*, by Shaun Tan. It is far less common in geographical research to analyze a single text than to analyze a succession or collage of texts, but Schnell demonstrates how even one text can be informative. The text in this case is so thoroughly dedicated to the theme of place attachment that it merits close examination. It is a wordless story so Schnell's explanation brings an extra layer of insight simply by re-telling the story in words. The analysis does not negate the value of a wordless text, however, because "Much of what happens to our narrator is beyond words" including sensations of fear, bewilderment, human connection and growing confidence "The imagery of the book ... gets us closer to the preverbal sense of place as it is directly experienced" (Schnell).

Johnson tacks between portrayals of Pitcairn Island in film, literature, popular magazines, news articles, the Internet, academic studies, a government memorandum, and nineteenth-century travel logs. These various texts (and interview findings) are unified by a focus on complementary aspects of place: perception, representation, attachment, spatiality, and temporality. While much of the source data could be called archival, the crucial distinction between this analysis and typical archival work is that representations are considered less as sources of information about the place than as information about how the place has been discursively constructed and understood throughout its history. Johnson's methodology points out tensions between the place's "infamous history and negative publicity" on the one hand, and the investment of Pitcairn Island with homey characteristics of safety and comfort by its residents, on the other hand.

Post takes us on a tour of memorial landscapes by drawing on online postings from TripAdvisor and Google Reviews. Surprisingly, a research method based on the "virtual" world of online experience and interaction offers substantial support to the idea that place attachment is embodied, suggesting that disembodied and embodied experiences can be mutually reinforcing. Visitors hold pieces of paper up to the polished stone wall at the Vietnam War Memorial in Washington, D.C. to take rubbings. They bring offerings. They walk down a slope toward the center of the memorial then back up again, performing a metaphorical journey into and out of war as experienced through the changing tension in muscles and bones. Also considering the Oklahoma City Memorial and the Good Deeds Chairs at Syracuse University, Post shows how an embodied relationship to the past as appraised through multiple sensory modalities, then shared in a digital form, can serve to guide subsequent embodied practices.

We pause to consider the power of toponyms before leaving the discussion of discourse analysis. The most specific form of discourse analysis considers the power of a place name, a toponym. Some place names are specific, like Kesennuma (Rowley) and Fountainbridge (Buckley), while others are generic, like wilderness (Olstad) and most beautiful village (Ducros). When a specific place is "labelized" by attaching it to a generic label, the labelers justify that labelization with yet more terms such as "quality," "worth," "value," "merit," and "appeal," and such symbolically freighted words take on a life of their own with economic, political, and architectural implications. Research methodologies must therefore treat place names as potent forces creating and shaping place attachment. Perhaps overstating the case, one of the interview respondents in Buckley's chapter argues:

"You rename a place and you destroy it." The match between place and name feels intensely natural and binding, but this is not actually the case, as is demonstrated by several of the chapters. Sometimes one place name is attached to multiple places, like the old and new Chipanga (Strong) or the three incarnations of Busch Stadium (Hurt). Conversely, one location can have multiple place names, like Virginia City a.k.a. Comstock (Barnett), Chacarita a.k.a. Barrio Ricardo Brugada (Smith), or Chipanga and the unfortunately titled "Unidade 6 of Bairro 25 de Setembro" (Strong). In such cases, it appears virtually inevitable that one of the names will be favored by those with strong attachments to the designated place while the other fades away or is used only by outsiders. To attend to the complexities of toponymy is a necessary element of place attachment research; the question of what to call places forms the point of departure for many studies of place attachment.

Conclusion

At the outset I prioritized methodological approaches to place attachment research: I set aside *what, who,* and *why* questions in order to focus on *how questions.* I based the above discussion on methodology on how people have studied place attachment. Now we can rejoin the broader field of inquiry about place attachment by returning to the *what, who,* and *why* questions.

In response to the question *"what* exactly is place attachment?" the contributors to this book have emphasized activities – the diverse expressions of agency that are paramount in place attachment. Whether in the hopeful rebuilding of a devastated city (Rowley), in replacing an obsolete baseball stadium (Hurt), in protecting a wilderness area from the impacts of throngs of visitors (Olstad), in protecting a slum from unaccompanied visitors (Smith), in reconnecting with nature by rafting down a river (Youngs), or in reconnecting with community by working in a garden (Buckley), place attachment is revealed through the diversity of human activities on and in the world. This work clearly challenges the common presumption that "local belonging is an anachronism" (Tomaney 2014, 507). Just as clearly, it casts into doubt the routine condemnation of place attachment on ethical grounds (Tomaney 2012). Merely to communicate is to intervene in place, setting off chains of cause and effect that lead to tangible forms of action, creating, sustaining, and destroying various aspects of places with complex and ambiguous ethical impacts (Peppler *et al.,* Ducros, Post, Olstad).

Moving to questions of *who,* it is inevitable that research into place attachment is structured around the distinction between insiders and outsiders. In Hurt's chapter the insider/outsider opposition is inherent to the game of baseball and crystalizes in the baseball stadium as "our place." In Strong's chapter, the residents of Bairro Chipanga are insiders distinguished by their collective loss of place. The protagonist in the graphic novel studied by Schnell begins as an outsider (an immigrant) who, over time, becomes an insider as he undergoes cultural adaptation and assimilation and as his family makes their home in their new country of residence. In Johnson's chapter, the insider and outsider divide was etched onto Pitcairn Island from the outset by an act of mutiny (breaking Us into Us and Them), but the status of "our place" has been reinvigorated by efforts at forced relocation and legal intervention from far away. In a few cases outsiders may even accentuate the qualities of place that are valued by insiders and the outsider-insider encounter may be mutually affirming, as Ducros finds, through the synergies created by preserving "our place" for their enjoyment (i.e., for tourists). The insider/outsider distinction evolves as places change and the "old guard" perpetually passes away, and finally as the outsiders who stick around slowly become insiders (Barnett; Schnell). The insider-outsider distinction therefore gives way to continuity across generations as each person identifies with others who have previously contributed their effort, energy, and commitment, and even their lives, to make a cherished place what it is (Ducros; Buckley; Johnson; Peppler *et al.*; Post). The divide between the living and the dead submerges into place attachment as people experience, enjoy, protect, maintain, preserve, restore, and rebuild a place that has been built by their predecessors, including those who are now gone (Rowley; Post; Johnson). Thus, the *who* of topophilia points to the creation of the "insider" as a powerful force for inclusion as well as exclusion, and topophilic defense of "our place" is ultimately a force of fusion.

Finally, taking up the question of *why*, our attention turns to the potential benefits that can arise from research that addresses place attachment. The primary reasons for doing scientific research in general are often classified as exploratory, descriptive, and explanatory (Yin 2011). Exploratory research lays the groundwork, identifying and defining an area of interest. This is research that puts a new place on the map. In regard to place attachment, this kind of descriptive research involves asking "where is it?" and "what is there?" In this volume only Strong and Smith provide examples of exploratory research. In contrast, explanatory research develops causal reasons for why things are the way they are and often depends on the generation of mathematical or symbolic models. Peppler *et al.* is the only chapter here that engages in significant explanatory research. Therefore, the contributors to this volume overwhelmingly emphasize the third approach, description. Almost all of the chapters in the book are markedly, and often quite artfully, descriptive. To adopt a cartographic metaphor that is particularly apt, descriptive research is about filling in the blank places on the map. Of course, the world map no longer has blank areas filled in with imaginary monsters and dragons. However, human feelings about place and place meanings remain largely uncharted. Sense of place is the remaining *terrae incognitae* (Wright 1947) and the contributors to this volume have done their best to describe this undescribed space.

References

Adams, P. C. 2017. Place. In *The International Encyclopedia of Geography: People, the Earth, Environment, and Technology*, edited by D. Richardson, 5073–5085. Hoboken, NJ: Wiley-Blackwell and the Association of American Geographers.

Adams, P. C., S. Hoelscher, and K. E. Till (eds.). 2001. *Textures of Place: Exploring Humanist Geographies*. Minneapolis: University of Minnesota Press.

Barnes, T. J. and J. S. Duncan. 1992. *Writing Worlds: Discourse, Text and Metaphor in the Representation of Landscape*. London: Routledge.

Cosgrove, D. and S. Daniels (eds.). 1988. *The Iconography of Landscape: Essays on the Symbolic Representation, Design and Use of Past Environments*. Cambridge: Cambridge University Press.

Crang, M. and N. Thrift (eds.). 2000. *Thinking Space*. London: Routledge.

Delyser, D. 2014a. Towards a Participatory Historical Geography: Archival Interventions, Volunteer Service, and Public Outreach in Research on Early Women Pilots. *Journal of Historical Geography* 46: 93–98.

Delyser, D. 2014b. Collecting, Kitsch and the Intimate Geographies of Social Memory: A Story of Archival Autoethnography. *Transactions of the Institute of British Geographers* 40: 209–222.

Dowling, R., K. Lloyd, and S. Suchet-Pearson. 2015. Qualitative Methods 1: Enriching the Interview. *Progress in Human Geography* 40: 679–686.

Duncan, J. S. and D. Ley. 1993. *Place/Culture/Representation*. London: Routledge.

Entrikin, J. N. 1991. *The Betweenness of Place*. Baltimore: The Johns Hopkins University Press.

Haraway, D. 1988. Situated Knowledges: The Science Question in Feminism and the Privilege of Partial Perspective. *Feminist Studies* 14: 575–599.

Jacobs, J. 1961. *The Death and Life of Great American Cities*. New York: Vintage.

Meinig, D. W. 1979. The Beholding Eye: Ten Versions of the Same Scene. In *The Interpretation of Ordinary Landscapes: Geographical Essays*, edited by D. W. Meinig and J. B. Jackson, 33–48. Oxford: Oxford University Press.

Pocock, D. 1981. *Humanistic Geography and Literature: Essays on the Experience of Place*. London: Croom Helm; Totowa, NJ: Barnes & Noble.

Pred, A. 1990. *Lost Words and Lost Worlds: Modernity and the Language of Everyday Life in Late Nineteenth-Century Stockholm*. Cambridge: Cambridge University Press.

Relph, E. 1976. *Place and Placelessness*. London: Pion Limited.

Sack, R. D. 1980. *Conceptions of Space in Social Thought: A Geographic Perspective*. Minneapolis: University of Minnesota Press.

Steacy, C. N., B. S. Williams, C. L. Pettersen, and H. E. Kurtz. 2016. Placing the "Analyst" in Discourse Analysis: Iteration, Emergence and Dialogicality as Situated Process. *The Professional Geographer* 68: 166–173.

Tomaney, J. 2012. Parochialism: A Defence. *Progress in Human Geography* 37: 658–672.

Tomaney, J. 2014. Region and Place II: Belonging. *Progress in Human Geography* 39: 507–516.

Tuan, Y.-F. 1974. *Topophilia: A Study of Environmental Perception, Attitudes, and Values*. Englewood Cliffs, NJ: Prentice-Hall, Inc.

Tuan, Y.-F. 1978. Literature and Geography: Implications for Geographical Research. In *Humanistic Geography: Prospects and Problems*, edited by D. Ley and M. S. Samuels, 194–206. Chicago: Maaroufa Press.

Winchester, H. P. M. 1999. Interviews and Questionnaires as Mixed Methods in Population Geography: The Case of Lone Fathers in Newcastle, Australia. *Professional Geographer* 51: 60–67.

Wright, J. K. 1947. Terrae Incognitae: The Place of Imagination in Geography. *Annals of the Association of American Geographers* 37: 1–15.

Yin, R. K. 2011. *Applications of Case Study Research*, 3rd ed. Thousand Oaks, CA: Sage Publications, Inc.

Index

Printed in the United States
by Baker & Taylor Publisher Services

Printed in the United States
by Baker & Taylor Publisher Services